I N T E R N A

T I O NAL

SPA CE

DES IGN

AW ARD

Idea-Tops 艾特奖

第七届艾特奖获奖作品年鉴

国际空间设计大奖艾特奖组委会　编著

湖南科学技术出版社

THE 「7」TH YEAR BOOK OF THE IDEA-TOPS

第七届艾

统筹策划：赵庆祥
英文编撰：陶心璐　王淑颖
设计总监：朱　辉
编辑委员：周党伟　陈　松
美术设计：容　荣
责任编辑：缪峥嵘
责任设计：殷　健

奖获奖作品年鉴

AWARD

CONTENTS 目录

Idea-Tops, The Most Internationalized and Professional Design Award in Mainland China

中国境内国际化程度最高的专业设计大奖

Idea-Tops艾特奖

国际空间设计大奖——Idea-Tops艾特奖，是中国境内国际化程度最高的专业设计大奖，建基于全球第二大经济体及迅猛发展的设计市场，旨在发掘和表彰在技术应用、艺术表现及文化特质再现上，具有创新意识的设计师和设计作品，打造全球最具思想性和影响力的设计大奖。

艾特奖为推动中西方设计交流搭建了一个沟通协作的平台，高水平的国际级评委、严谨公正的评奖机制及奖项设置……使艾特奖成为中国境内最具国际化和专业性的设计奖项，也成为了世界设计业了解中国设计的一个窗口。

艾特奖参与者可谓众星云集，包括全球三大设计事务所之一的Gensler设计总监Graeme Scannell、"中国第一高塔"广州塔设计者Mark Hemel，全球酒店设计公司五强、BBG-BBGM建筑与室内设计公司设计董事Robert J.Gdowski，拥有140年历史的国际知名建筑事务所Woods bagot全球总监Rodger Dalling，英国首相官邸——唐宁街10号设计者、BBC苏格兰总部设计师Ross Hunter，希尔顿国际酒店集团主创设计师Martin Hawthornthwaite，曼联俱乐部亚太区首席设计师Mike Atkin，深圳机场T3航站楼设计者Fuksas夫妇、2008年普利兹克奖获奖者、法国当代著名建筑师让•努维尔(Jean Nouvel)等。此外，2015年米兰世博会11大国家馆强势竞逐2015艾特奖，2016年威尼斯双年展14国顶尖级参展设计师携作品参与角逐2016艾特奖。

原于东方，面向世界，Idea-Tops艾特奖崇尚的，是设计师永不枯竭的智慧与前瞻性的革新思想，以及他们审美和工艺上均显卓越的设计作品。恪守专业、严谨、公平、公正的原则，Idea-Tops艾特奖以较高的专业标准、专业发展、专业责任以及专业沟通来促进设计业的发展，每届艾特奖均邀请全球设计领域资深专家、学者、顶尖建筑师和设计师、知名人士、财经专家、意见领袖以及有影响力媒体担纲评委。

作为表彰建筑和室内设计界杰出人才的重要奖项，获得艾特奖的肯定也就是向全球昭示了精英们在建筑和室内设计界的顶级荣誉。在中国，超过2/3的知名房地产开发商将艾特奖获得者锁定为首选合作伙伴。在欧美、亚太地区，艾特奖正逐步成为设

International Space Design Award—Idea-Tops is the most internationalized and professional design award in Mainland China. Its establishment was based on the rapidly developing design market of the world's second-largest economy. Idea-Tops aims to discover and praise designers and design works that are innovative in the space form, technology application, art performance and culture presentation, and create the most thoughtful and influential design award around the world.

Idea-Tops builds a platform for communication and collaboration between Chinese and Western designs. The high level of international judges, rigorous and impartial mechanism of awards and the setting of award categories make Idea-Tops not only become the most internationalized and professional design award in Mainland China, but also a window for the world design industry to understand Chinese design.

Lots of masters participated in Idea-Tops, including Graeme Scannell (design director of Gensler—one of the world's top three design studios), Mark Hemel (designer of Guangzhou Tower—China's tallest Tower), Robert J.Gdowski (designer principal of BBG-BBGM—one of the world's top five hotel design companies), Rodger Dalling (global director of Woods Bagot—the international renowned architecture firm which has a history of 140 years), Ross Hunter (designer of No.10 Downing Street and BBC Scotland headquarters), Martin Hawthornthwaite (chief designer of Hilton Hotels Corporation), Mike Atkin (Asia-Pacific principal designer of Manchester United Football Club), Massimiliano and Doriana Fuksas (designers of Shenzhen T3 Airport), Jean Nouvel (winner of 2008 Pritzker Architecture Prize, famous contemporary architect in French), etc. In 2015, 11 national pavilions from Expo Milano 2015 competed in Idea-Tops, such as Italy pavilion, Turkey pavilion, Austria pavilion etc. And in 2016, many designers who have exhibited projects at the 2016 Venice Architecture Biennale competed in Idea-Tops. These designers are from 14 countries, such as Italy, UK, France, Germany, Canada, Israel, Belgium, etc.

Idea-Tops originated from the East and faces to the world. What it underscores are designers' inexhaustible wisdom and forward-thinking innovation as well as their works which are excellent in aesthetics and craft. With the principles of being professional, rigorous, impartial and fair, Idea-Tops promotes the development of design industry by professional standards, professional development, professional responsibility and professional communication. In each session, Idea-Tops invites a fantastic and distinctive judge panel, including senior experts, scholars, top architects and designers, celebrities as well as financial experts, opinion leaders and influential media representatives in the global design field.

Idea-Tops is an important award which praises outstanding designs. Getting its recognition is a global announcement of designers' great honor in the architectural and interior design field. In China, over two-thirds of renowned real estate developers will choose Idea-Tops awardees as preferred partners. Idea-Tops is becoming a hot topic in the design community in Europe, America, Asia

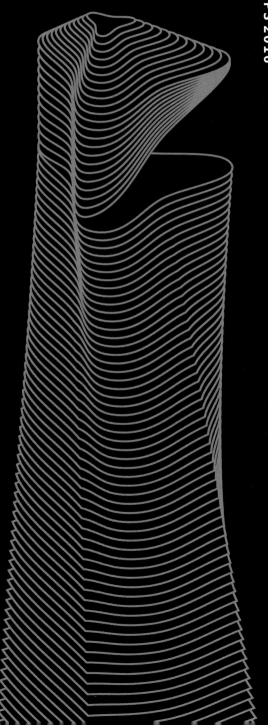

向设计致敬

他们从平凡中来，
又从平凡中创造伟大。
他们不遵循规则，
却重新定义了规则。
他们用变革精神，
为我们的生命注入色彩。
我们尊敬并感谢他们
向设计致敬！
向设计师致敬！

IDEA-TOPS 2016

序言 P

成长、发酵、裂变，我们永不止步
We Grow, Develop, Change and Never Stop

赵庆祥 / 艾特奖组委会执行主席 、 深圳市政协委员
Zhao Qingxiang, Executive Chairman of Idea-Tops
Organizing Committee, Member of CPPCC in Shenzhen

艾特奖始创于2010年，迄今已走过7个年头。全球化时代的浪潮深刻影响着人类社会的发展，过去7年，中国经济也获得了举世瞩目的成就并且发展成为全球第二大经济体。正是在这种背景下，艾特奖逐步成长、发酵并且裂变成一个属于全球设计界的国际性平台。从另一方面讲，艾特奖已不仅仅是一个奖项，发掘和培育最佳设计师，传播先进设计文化、推动设计业的交流、合作与进步，已成为这个国际性平台的重要属性！

2016年，艾特奖先后走过50多座城市，举办了77场大型的设计推广论坛，50多场高规格的展览与设计沙龙，20多万名设计师参与。

2016年，艾特奖陆续在全国建立了50多个城市推广中心，覆盖设计人群30多万人次，全面推广艾特奖；与此同时，艾特奖还与拉丁美洲的国家与地区以及俄罗斯、日本等国家专业机构建立合作关系，让海外设计师了解并参与艾特奖，让艾特奖逐步在全球落地。

2016年，艾特奖得到了中国科学院院士郑时龄先生、中国工程院院士崔愷先生、中国工程院院士孟建民先生、亚洲大学刘育东教授以及海内外顶级专家、设计师和学者们的鼎力支持，他们赋予了艾特奖更高的格局、更前端的智慧以及更鲜明的发展方向。当然，还有近百万参与或关注艾特奖的设计师朋友们！你们的关注与支持，赋予了艾特奖强大的生命力。

2016年，艾特奖共收到来自35个国家和地区的设计师参赛作品6020件，其中更有14位威尼斯国际建筑双年展参展设计师参与角逐。本届艾特奖参赛作品的数量、质量都创了历届新高。

今天，《第七届艾特奖获奖作品年鉴》全球首发，真的，非常有分量！在此我要向他们、向所有获奖设计师们表示祝贺！

It has been 7 years since Idea-Tops established in 2010. In the past 7 years, the globalization has deeply affected the development of human society, and Chinese economy has become the second largest in the world. Just under such background, Idea-Tops grows, develops and changes into an international platform for the global design community. Idea-Tops is not only an award, but an important platform that discovers and cultivates best designers, spreads advanced design cultures, promotes communication, cooperation and progress of the design industry.

In 2016, Idea-Tops held 77 large design forums and over 50 high quality exhibitions and salons in more than 50 cities, which have attracted more than 200,000 designers.

In 2016, more than 50 city promotion centers have been established to comprehensively promote Idea-Tops in China, which has covered over 300,000 designers. Meanwhile, Idea-Tops has built partnership with professional agencies in Russia, Japan, Latin America, etc. enabling foreign designers to know and participate in Idea-Tops.

In 2016, Idea-Tops gained support from Zheng Shiling (Academician of Chinese Academy of Sciences), Cui Kai (Academician of Chinese Academy of Engineering), Meng Jianmin (Academician of Chinese Academy of Engineering), Liu Yu-Tung (Vice President of Asia University, Taiwan) and other domestic and overseas experts, designers and scholars. They increased the level of Idea-Tops, enriched wisdom and guided its development. Of course, the attention and support of millions of designers have given Idea-Tops great vitality.

In 2016, Idea-Tops received 6020 entries from 35 countries and regions, including 14 projects from the Venice Architecture Biennale. The quantity and quality have created new records.

Today, the Award-Winning Works of Idea-Tops 2016 is published for the first time. It is very valuable and we congratulate all the winners!

序言

文化传承是设计创新的基点
Cultural Inheritance is the Basis of Design Innovation

崔愷

院士/2016艾特奖国际评审委员会主席
Cui Kai, Academician of Chinese Academy of Engineering,
Chairman of Idea-Tops International Jury 2016

艾特奖是一个涵盖了建筑设计与室内设计的跨界奖项，也是一个很好的交流平台，它既增进了中国建筑设计界的相互了解，又有利于提升国际上对中国设计界的认识。

本人第一次参加艾特奖的评审，对比之前参与的其他设计奖项，我觉得艾特奖非常好的一点是它的跨界性：建筑设计、室内设计、环艺设计、照明设计，甚至不同类型与风格的项目都能参加评奖，参赛阵容很强。2016年有6000多幅作品参赛，这是巨大的工作量，对评审工作也是一个挑战。终评我们评的是提名奖作品：18个类别，每个类别5套提名作品，总共90套作品，是经过前面几轮评审层层筛选出来的。这90套作品非常优秀，里面有很多海外团队的项目，这说明艾特奖在国际上已经具备一定的影响力。

作为评委，我认为一个好的作品首要先创新。任何一个建筑或室内设计作品，我们比较关注它是否和以往这类项目的惯常设计有所不同。其次，它应该比较积极地响应环境和生态，所以我很看重这里面具有绿色设计取向的作品。最后，它还要很好地传承文化。这次有不少作品中都能看到有某种地域文化基因，它们体现了设计师对自己的文化与艺术的认同和传承的意愿。我觉得这个非常好，文化的传承一直是设计创新中非常重要的基点，如果没有这个基点，我们仅仅讲未来、讲手法、讲技术，未必能得到大众的认可。其实我特别想看建筑和环境之间的那种互动关系，特别喜欢有创意，同时又和周边环境融洽得非常好的设计，这种融合不仅仅是建筑形象，或者空间衔接上的关系，还包括地方气候特点和现实生活之间的关系。

比方说今年的公共建筑奖，我印象特别深的是新加坡的国立美术馆改扩建项目——设计师把有东南亚乡土气息的竹簾棚架巧妙地嵌入到当地经典而庄重的欧式历史建筑中，你一下就被那种形式的对比和文化的和谐性所吸引，还有对气候环境的应对，非常符合新加坡的特色，手法很简洁，设计很精巧，整体非常棒。还有另一个高尚别墅设计也很有正能量，设计师没有用传统豪华装饰的手法去显示主人的财富，而是设计了一个充满趣味的立体花园作为别墅空间的中心，在周围豪华别墅群中独树一帜，又和周边的绿林融为一体，不仅体现出生活价值观的变化，也是对生态城市的引导，非常好！当然这不仅是设计的好，更重要的是主人的价值观使然，给我们很多启发。所以说参加一次高水平的评奖也是一次很好的学习机会，其间与其他评委的互动讨论也颇有收获。

期待未来在艾特奖这个平台可以看到更多具有地域文化基因和生态绿色价值观的创新设计，也建议艾特奖进一步开放设计领域当中的互动性以增强设计界的跨界交流，坚持做好专业权威的评审工作，为这个行业创造更长远的价值。

Idea-Tops is a cross-border award that covers architectural design and interior design, and a very good platform for communication. It not only enhances the mutual understanding of Chinese architectural design, but also helps to raise the awareness of Chinese design community internationally.

This is my first time to participate in the review of Idea-Tops. Compared to other design awards I took part before, I think it's very good that it crosses different fields like architectural design, interior design, environmental art design and lighting design. Entries can have different types and styles, so the competition is very fierce. There were more than 6,000 entries in 2016, such huge workload was a challenge for the judges. In the final review, we have evaluated 90 nominated works: 5 projects in each of the 18 categories, and all were selected after several rounds of reviews. Those projects were excellent, and many of them were from foreign teams, implicating that Idea-Tops already has some international influence.

As a judge, I think firstly, a good work must be innovative. For any architectural or interior design work, we are more concerned about whether it is different from the usual design. Secondly, it should be active in responding to the environment and ecology, so I value those works that have green design orientation. Finally, it needs to inherit culture well. This time we saw certain regional cultural genes from some works, they reflect the designers' recognition of their own culture and art, and their will to inherit them. I think this is very good since cultural heritage has always been a very important basis of design innovation, if without it we would only talk about the future, skill and technology, which may not be recognized by the public. I particularly want to see the interaction between architecture and environment, especially love creative design that is in perfect harmony with the surrounding environment, this integration is not just an architectural image, or a relationship on spatial connection, but it also includes the relationship between local climate characteristics and real life. For example, this year's winning project of the Best Public Architecture Design Award—the reconstruction of Singapore National Gallery, I was very impressed by it since the designer cleverly embedded Southeast Asia bamboo scaffolding into the classic and solemn European style historical architecture in the local area. you would be attracted as soon as you feel the contrast and cultural harmony. Furthermore, the response to the climate environment is totally accord with the characteristics of Singapore, moreover, the technique is simple and the design is very exquisite, so the whole project is really amazing. There was another noble villa design that made people feel very positive. Without using traditional luxurious decoration to show off the owner's wealth, the designer created a truly interesting vertical garden in the center of the villa, making it very unique among the surrounding luxury villas yet it's integrated into the green forest. The villa not only reflects the change of life values, but also guides us to the ecological city, and this is very good! Of course this is not merely about a good design, it is the owner's value that is more important and gives us a lot of inspirations. Therefore, participating in the review of a high level award is also a good opportunity to learn, and the interaction with other judges makes me feel fruitful.

I hope that, through the platform of Idea-Tops, we can see more creative designs that show regional cultural genes and ecological and green values. I also suggest Idea-Tops to open interaction in the design field so as to enhance the cross-board communication among the design community, insist on professional and authoritative reviews, and create more profound values in this industry.

2016

Idea-Tops艾特奖
评审委员会

IDEA-TOPS
JURY PANEL 2016

¹ 崔愷
中国工程院院士、中国建筑设计研究院总建筑师

² 孟建民
中国工程院院士、全国建筑设计大师

³ 刘育东
台湾亚洲大学副校长、哈佛大学建筑学博士

⁴ 梁景华
香港十大设计师之一、PAL设计事务所创办人及首席设计师

⁵ MICHELE MOLE
米兰世博会国家馆设计师、意大利共和国艺术文化总统勋章获得者

① ②

③ ④

⑤

Idea-Tops in the Eyes of Designers

设计师们
眼中的艾特奖

2016年12月6日，2016年度艾特奖系列活动在深圳市龙岗区中海凯骊酒店盛大举行。在室内设计行业迅猛发展、机遇与挑战并存之际，2016艾特奖吸引了全球设计的眼光，共同探讨设计、共话设计未来，形成了强大的磁场力和向心力。一个设计大奖为何能形成如此巨大的影响，掀起行业飓风？艾特奖本身有哪些魅力？在国内外设计大师及专家学者眼中，艾特奖究竟是什么样的？让我们走近国际顶尖设计大师，感受艾特奖的真实魅力。

On December 6th, 2016, the serial activities of Idea-Tops took place at Coli Hotel in Longgang District, Shenzhen. At the moment when opportunities and challenges coexist in the booming design industry, Idea-Tops captured the global design community's attention. People gathered in Shenzhen to discuss design and future trends, which brought dynamic influence. Why does Idea-Tops make such a great influence on the industry? What characteristics does it have? What do domestic and overseas design masters, experts and scholars think of Idea-Tops? Let's take a close look and find out the charms of Idea-Tops.

Alireza Razavi

（法国 / France）
Studio Razavi建筑事务所创始人
Founder of Studio Razavi Architecture

很高兴能来到这，因为中国具有非常古老的文化，我们可以从中学到很多东西。同时，中国有它自身的发展道路，并探索出了一系列独特的解决方案，在这里我感到非常开心。当下有许多建筑趋势，其中有一种趋势非常注重几何结构，对于我自己来说，我对建筑物是如何建造的比较感兴趣，这种趋势可能更关注于结构和物质性。

我去过北京、西安和深圳。我非常喜欢北京，它有历史建筑，也有非常现代的建筑。深圳给来自欧洲或美国的人留下很深刻的印象，30年前这种情况应该比较少，我希望我能有更多时间在这里度过。

It was a pleasure being here, because China has a very ancient culture, so we have a lot to learn from that, and also China has had a very specific path of its own and has come up with very specific solutions. I'm very happy to be here. There are many architectural trends these days, and one that emphasizes geometry a lot, I think I'm more interested in the ones that show how buildings are built, so emphasize more maybe on structure and materiality. I have been to Beijing, I have been to Xi'an and Shenzhen. Beijing, I liked it very much, both for the historical architecture and the very recent architecture. Shenzhen is very impressive for someone coming from Europe or the US, because I believe there wasn't much 30 years ago. It's very impressive and it works very well, I wish I could spend more time here.

Maria Alessandra Segantini

（意大利 / Italy）
美国麻省理工学院客座教授，C+S Architects
建筑公司总监及合伙人
Visiting Professor at MIT, one of the directors and partners of C+S Architects

我很荣幸能够参加这个盛会，我认为建筑师和创意群体在一个国际性的平台上进行交流、分享是非常重要的。同样，这个采访也很重要，它让我们的思想在不同人群和社区之间得到传播，能够激励我们的工作。如今我们面临各种各样的危机，如环境危机、社会的不平等、移民问题等，此外，在世界各个地方有非常多不同的传统。因此，我想说那些能够解决危机，或在紧要关头能让问题变得更简单的设计才是有价值的。显而易见，建筑师是形态的创造者，因此我们应该设计出美观、强大、可持续的空间，让不同的群体快乐地生活和工作。

我觉得独特性比品牌更加重要，我对那些仅仅在形式上具有独特性，即只有品牌的设计公司持怀疑态度，我认为如果我们能将品牌和内容结合起来，并且如果内容逐渐变成品牌，那么我们可以创造一个更加美好的世界。

I was really honored to be a part of this event. I think it's always very important for architects and creative communities to exchange ideas, opinions and to share that at an international level. And the important thing is also this interview, because it allows these ideas to be spread among people and communities, which empower our work. So in contemporary, we do have a lot of urgencies, and these urgencies are environmental urgencies, we have social inequalities, we have migration, we have lots of heritage in many places of the world, so I would say that all design which really tackles the urgencies and takes a position to fix the complexities of our crucial moment when we leave this ease the design which values. But obviously, architects are form makers, so we need to design spaces which are so beautiful, powerful and sustainable to allow communities to live, work, and be happy in it.

It's important to have an identity more than a brand, but you know I'm a little bit suspicious on those design firms who have only a formal identity, so only brand, I think if we can combine a brand with content, and if content becomes a brand, then we can make a better world.

Chika Muto

（日本 / Japan）

艾特奖日本合作伙伴、Neoplus Sixten创始人、建筑设计品牌咨询与公关推广专家

Partner of Idea-Tops in Japan, Founder of Neoplus Sixten, senior PR&Communication Specialist in the architecture and design fields

在东日本大地震之后，许多日本建筑师和设计师将他们的思维转向更多的领域，他们建造了社区空间和场所让人们能够聚集在一起。因此我认为，通过这些项目我们可以学到的是，要建造一些人类真正需要的东西，而不仅仅是建造更多的东西。对于日本及其他周边国家的人们来说，艾特奖是进入中国市场的一个非常好的渠道，我希望更多的设计和理念能够通过这个平台进行交流。

After the Great East Earthquake in Japan, many Japanese architects and designers have shifted their minds towards more, creating community spaces and places where people can gather, so I think that through these projects we could learn how to not to create more and more, but to create something that's really necessary for human beings. Idea-Tops is a great gateway for Japanese people to enter into Chinese market, as well as other neighboring countries, Asian countries, and I hope that more designs and ideas will be exchanged through this platform.

Xavier Lagurgue

（法国 / France）

法国巴黎尔德塞纳国立高等建筑学院教授、环境质量的专家、XLGDL国际建筑事务所联合创始人

Professor of École Nationale Supérieure d'Architecture de Paris–Belleville, Specialist of environment quality, Founder and Partner of the Franco-Austrian agency XLGD architectures

我对主办方表示祝贺，我想这对中国建筑发展来说是件非常好的事情。能来参加我感到很自豪，谢谢。我觉得环境质量在城市中呈现出前所未有的重要性，因为我们需要生态系统服务，需要他们来获得新鲜的空气、水、食物以及主要的材料，如果我们没有这些东西，我们就无法在城市中生存。因此，我认为建筑师、设计师和城市规划师需要非常谨慎，且责任重大，这一领域是开放的，它让我们试图寻找新的形式和方法为所有的生物建造新的空间，不仅仅是人类，还包括植物和动物。对于我来说，很重要的一点是不再把人类和非人类分开，而是应该同时考虑到两个方面，将二者真正地联系在一起。动物和植物不是物体，而是生物，它们有自身的作用，因此也应该有适合自己的独特空间，这对于未来的建筑师和设计师来说是一个巨大的挑战。

I congratulate the organization, it's a really good thing I suppose for China's architectural evolution. And I'm very proud to be here, and I say thank you. I think, you know, more than ever, environmental quality is important in the city, because we need something which is called ecosystem services, we need them for air fresh, water, for food, for ruling materials, and if we don't have them, in a moment it would be impossible to live in our city. So I think architects, designers and urban planners have got quite responsibilities in discretion, and the field is open for us to try to find new forms and new ways to set new place for all the living creatures, not only human kind, but all the living creatures, vegetations and animals. You know, the principal thing for me, is not to separate any more, human and non-human, we need to consider both together, really linked. And so animals and vegetations are not objects, they are living creatures, they have got some functions, and so it's necessary for them to conceive space, adaptive space specially for them, that's a great challenge for new architects and new designers in the future.

Giuseppe Strappa

（意大利 / Italy）
罗马大学建筑与城市设计学院教授、建筑设计系博士项目协调员
Professor of Architecture Faculty of Sapienza –
Universityof Roma, Italy, Coordinator of PhD program
of architecture design

我觉得如今我们不能谈论建筑或城市转型的方向，
这是提出问题的一种抽象的方式。不同的文化和价
值观里有不同的方向，我们必须认真对待这些差异
和特性，因为全球化并不是一个把什么东西都混合
在一起的大熔炉。全球化的重要性是不同特性之间
的辩证关系，因此我们必须重视这些差异。
相对来说，我比较了解意大利的城市，每个城市都
有其自身的问题。我觉得其中最相关的一个问题是
城市文化的混合。我们有非常非常多的移民，有人
认为解决问题的方法是阻止移民，但是这是一个历
史现象，我们不可能压迫成千上万到欧洲来的人。
这会改变我们思考问题的方式，我们必须重新思考
我们的文化，甚至是我们设计城市的方法，我觉得
这是其中一个最为重要的问题。
另外一点就是我们在意大利城市解决两个重大问题
的方式，它们分别是内部城区以及边缘区域的扩
张。这是两个具有不同特性的问题。内部城区大部
分是一些历史悠久的区域，在我看来，不得不对它
们进行改造。我们保留了这种文化遗产，并且必须
保护这一遗产，让其不被侵犯。但我有一个不同的
观点，城市、甚至是古老的城市就是一个城市生命
体，和其他生命体一样，它必须通过发展和改变才
能生存下去。

I think that nowadays we can't spea
direction in architecture or urban tra
This is an abstract way of posing the
There are, at present, many directio
belong to different cultures, differen
we have to take care of these differe
specificities. Globalization is not a m
which everything is mixed. The impo
globalization is the dialectic relation
differences, so we must care these d
To come to the question, the cities w
better, of course are the Italian ones
specific problems. I think one of the
relevant problems is the hybridizatio
urban culture. We have a very stron
immigration. Someone think that we
the problem arresting the immigratio
phenomenon is a historic one. It's n
let millions of people pressing to arr
Europe. This will change much our w
thinking. We must rethink of our cult
of even designing the city. This is on
problems in my opinion mostimporta
The other one is the way we manage
important issues we have in Italian c
city and the expanding periphery. Th
problems which have got their speci
of the inner city are historical, the hi
which, in my opinion, have to be tran
We have a culture of preservation, o
We must preserve this heritage, and
be touched. I have a different opinio
even the old city, is an urban organis
any organism, it must develop and c
survive.

Jiarong Goh

（新加坡 / Singapore）
studioMilou助理建筑师
Architectural Associate of studioMilou

我们感到非常荣幸，非常感谢主办方和评委对我们
两个项目的肯定。我代表我们的首席建筑师Jean-
Francois Milou表示歉意，他很遗憾不能来参加，
但是他真的很想来，再次感谢给予我们这个奖项。
第一个项目是新加坡的一个住宅设计，对我们来说
它很特别，因为这是我们在新加坡的第一个项目。
studioMilou实际上是法国巴黎的一个建筑事务
所，但在做另外一个获奖项目，即新加坡国家美术
馆的时候，我们搬到新加坡并在那里也设立了公
司。
新加坡国家美术馆这个项目保护性地利用了两座非
常重要的历史遗迹，我们将它们合二为一并变成了
一个重要的艺术场馆。项目非常注重对遗迹的保
护，不仅保存了新加坡的艺术，还在建筑中营造出
美妙的光线。

We're very honored and thank you very much
for the organizer and jury for giving
recognition to these two projects. On behalf
of my principal architect Jean- Francois

Milou, I'd like to express he's re
for not being able to attend, bu
wanted to come, so we thank yo
giving us this award.
The first project is a residential
Singapore, to us it's a very spe
project, because it is the first re
project we did in Singapore. stu
is actually a firm from France, fr
and because of the next project
won, which is National Gallery S
we actually shifted to Singapore
have set up a company there.
National Gallery Singapore is ac
another project that there is
conservational and adaptive us
very important monuments in S
the design is actually to conver
two monuments, and join them
to become an art gallery, a very
important gallery in Singapore,
project itself is all about the ser
conservation and also creating
spectacular light within the buil
itself, to retain the art of Singap

Pier Paolo Tamburelli

（意大利 / Italy）

Baukuh建筑事务所创始人之一、米兰理工大学客座教授
Co–Founder of Baukuh, Visiting Professor at the Milan Politecnico

我因"回忆之屋"获得这个大奖，这个建筑是米兰的一个档案馆，相对来说面积较小，只有2500平方米，但是它非常重要，因为它是这个城市的纪念碑。实际上，这栋建筑收藏了第二次世界大战时期游击队员抵抗法西斯主义的相关文件。它对这个城市以及整个国家具有重大意义，当然，要在米兰市中心建造这样一栋建筑，我们的责任非常重大，因为你知道这个区域非常古老，因此很难对其进行改造。我们想要建造一个简单但庄严的空间，用于保存及展示那些重要的文件，并且这个空间不仅仅是对历史学家或相关领域的专家开放，小孩、学校及其他群体也能够轻松地接触到这些文件。

I received this award for the House of Memory, that is an archive in Milan. It is a relatively small building, 2500 square meters, but it is important, because it is a monument for the city. The building in fact contains the collection of the documents produced by the partisans fighting against Fascism during World War II. It is a very important archive, important for the history of the city and of the country, and of course it was quite a responsibility to deal with it and to build this building in the city center of Milan, that as you know, is a rather old part of the city that is difficult to modify. We wanted to provide a simple and dignified place where it could be possible to preserve and expose these important documents, and where the documents could be easily accessible for an audience that is not necessarily made only of historians or experts on the subject, but also kids, schools and so on.

Michele Molè

（意大利 / Italy）

Nemesi公司的创始人、总裁，意大利米兰世博会国家馆设计师
Founder and Director of Nemesi, Designer of Italy Pavilion Expo 2015

世界瞬息万变，科技发展也非常迅速。科技并不是我们的救世主，但它是创造新事物，有情感的事物，创造新的城市、空间和建筑，从而获得更好的生活基础。从这一点来看，科技为我们带来新的挑战和重要的目标。例如我们可以利用新科技创造充满创意和个性化的建筑。另一方面，科技引导我们不能污染环境，摧毁世界。建筑也和人类一样需要呼吸，需要和其他空间交流，我们应该让建筑和景观之间产生互动，而新科技给予了我们这样的机会。

我感觉非常好，这里有很多来自当地和其他国家的设计师，他们很有趣也非常优秀。我参加了这次的评审并和其他评委一起选出了18个类别的获胜者，我们觉得这些项目的水平和质量都很高，这一点很重要。另外，我认为很有趣的是近几年来发生了一些变化，现在中国建筑师和设计师的水平都很高，因此我们之间有了交流，而不再仅仅是教或学。全世界超过45%的正在建设的建筑都在中国，中国就像是一个不可思议的世界性实验室，我们在其中学习如何发展城市和建筑，而艾特奖的论坛是思考这些问题的非常重要的一个平台。

The world right now runs really fast, and the technology changes fast. The technology is not the saver of our world inside lonely technology, but it's the base to create something new, something that have emotions, to create new city, new space, new building for higher level of life. Technology, from this point of view, offers to us the new challenge, new really important objective. For example, with the new technology, we can decide the building this moment, create something really innovative and personalized. And from other side, the technology gives the instructs to us that don't destroy our world, our environment with pollution, with other elements, but to create a dialogue between building and environment, the building have to breath like people, have to exchange with another space, and create interaction between building and landscape, and the new technology gives us this kind of opportunity. My feeling is good, quite good, because there are a lot of interesting and really good international designers and Chinese designers here. I took part to the jury of the award, and I selected winners of the 18 categories, I can say that the level, the quality I found, and all the other judges found in the projects were really high, this is very important from my point of view. Another aspect is that in these last years, something is changed, in this moment Chinese architects and designers have a really good high level, it's not like some years ago, and this is for me, a really interesting element, because right now there is an exchange between us, it's not only teaching, or learning, but it's an exchange, this is really interesting. You have to know that China in this moment, it's like a worldwide laboratory, because more than 45% of the buildings that we are realizing all around the world right now are realized in China, so it's a really incredible laboratory in which we are learning how to develop city, how to develop new building, new architecture, so I think the Idea-Tops forum is really important to think about these issues

Chris Mitchell

（美国 / US）
M+M创意工作室创始人和CEO
Founder and CEO of M+M Creative
Studio

这次能够获奖对于我们来说太不可思议了，因为我们以前从来没接触过这个奖项，当我们知道这个奖项时，我便产生了兴趣，能够参与到这个奖项已经让我们很惊喜，最终获得大奖真是太酷了！我不知道该怎么表达，因为我太兴奋了！这个设计想要呈现的是一种未来感，这是为洛杉矶的一个新办公总部所做的设计，公司有600名员工，所以我们想要创造一个能够鼓舞人心，且让员工精力充沛的办公空间，让他们愿意待在办公室里尽自己最大的努力工作，让其成为一个相互协作，具有团队精神的空间。能获得大奖让我们非常惊讶，但事实上是我们的客户获得了这个奖，这一点让我们所有人都非常高兴，谢谢！

This is amazing for us, I mean, this is something that we never had exposure to, and when we found this competition, I got interested in this competition, you know, just being a part of it was amazing, but winning it is extra really cool. I almost don't know what to say, cause I'm so excited about it. So the design is meant to be futuristic, but this is a design for a new headquarters in Los Angeles, and this is a design for 600 people, so it's really meant to invigorate, to energize, to bring the people into the work space, and to make them wanna be there during the day, you know, almost every day frankly, and to really put forth their best effort in the space, making it a collaborative space, a teaming space. And the client, I mean, for us really, the fact we won the award is quite amazing, but the best part is the client won the award, and that's what really makes us all happy, so thank you very much.

Claudia Pasquero

（英国 / UK）
ecoLogic Studio公司联合创始人
Co-Founder of ecoLogic Studio

我很荣幸获得这个奖，并且非常享受这个盛会，尤其是非常享受研究者和实践者之间的交流，让我受益匪浅，这一点并不常见，因此我觉得非常好。我对生态学感兴趣，但通过竞赛和数字设计来进行研究。我们看到与自然系统进行互动的必要性，例如通过城市中存在的不同元素，我们该如何减少空气污染，而不仅仅是净化空气，通过与空气的相互作用，创造自然系统，就不仅仅是输出，还会有输入，有更多的新陈代谢，就像有机体一样。

I'm really honored of the prize, and I really enjoyed the event, especially I really enjoyed the interaction between researcher and practitioner, I think that was very fruitful, and you cannot find this very often, but that was really nice I think. I will be interested in working on ecology, but working with the ecology through competition and digital design. And we see the necessary of interacting with natural system, so for example, with elements that are present in the city, with air in the city and the pollution, how can we eliminate to pollute it, but not by simply clean it, but by interacting with it, by creating our natural system, there'll not only have output, but there are also have input, so there are more metabolic, like organism.

Chang Yong Ter

（新加坡 / Singapore）
Chang建筑事务所创始人和设计总监
Founder and Director of CHANG Architects

我感到非常激动，因为身为一个华人建筑师能够得到这个国际奖项，我感到非常荣幸和鼓舞。这个别墅是为一个四代同堂而设计的，顾客有3个孩子，为了打算他们将来结婚、成家，可以留下和他们一起住。因为现在新加坡生活大部分是小家庭，很少三代同堂，所以这是一个很难得的机会，可以设计一个屋子让三代同堂回来一起聚一聚。

I'm so excited, as a Chinese Singapore architect, I feel very honored and inspired to win this award. This house is for four generations. The client has three children, and he planned that they will live with the parents after marriage. Most of the families in Singapore are small, very few are of three generations, so this is a very rare opportunity for me to design a house for the generations to come back and spend time together.

Alida Cornelia Brouwer

（荷兰 / Netherlands）
罗斯加德工作室展览&商务拓展总监
Head of Exhibition & Business Development of Studio Roosegaarde

我们在荷兰的东部设计了Waterlicht这个项目，这是为了展示如果没有堤坝那里将会变成什么样子。我们的任务与阿姆斯特丹国立博物馆相关，博物馆购买了一幅由Jan Asselijn在1651创作的画作 "The Breach of the Saint Anthony's Dike near Amsterdam" ，它是关于阿姆斯特丹附近圣安东尼堤坝的决堤。如今也有很多堤坝遭到破坏，如果全球持续变暖，海平面继续上升，那么会有越来越多的堤坝决堤，于是我们将这一点通过Waterlicht展现出来，吸引了很多人前去参观。因为有堤坝，水面到达一定高度时人们无法看到，但通过光线，我们将水的形态呈现

painting "The Breach of the Saint Anthony's Dike near Amsterdam", painted by Jan Asselijn in 1651, which is about the dike breakthrough in the Netherlands. And dikes breakthrough still happen there days, and certainly they will happen if the global warming will go on and the sea level will rise, and we documented it, we represented that with Waterlicht, so a lot of people go to experience how it look like. We wanted to show the people what they don't see, the water would be this high, but you don't see it because of the dikes, but with lines of light we could visualize how high it would be, so people could

Idea-Tops Awarding Ceremony 2016

2016年度
艾特奖颁奖盛典

16年12月6日，蜚声中外的国际空间设计大奖——2016第七届艾特奖颁奖典礼在
设计之都"深圳市龙岗区中海凯骊酒店盛大举行。来自中、美、英、法、意大利、
罗斯、日本、希腊、奥地利等35个国家和地区的1500多名设计师代表、知名企业
出席本届艾特奖颁奖典礼。

16年艾特奖共收到来自中国、美国、意大利、德国、荷兰、葡萄牙、比利时、日
澳大利亚、西班牙、奥地利、希腊、俄罗斯、墨西哥、印度、罗马尼亚、以色
巴西等35个国家和地区的建筑与室内设计师参赛作品6020件，2016年全球最具
响力的威尼斯建筑双年展，包括英国、奥地利、以色列、埃及、尼日利亚、亚美尼
塞尔维亚、意大利、法国、德国、芬兰等14国顶尖级建筑师也携作品参与本届艾
奖的角逐，一场"国与国"之间的设计较量再次燃爆本届艾特奖年度盛会。

港卫视、深圳卫视、南方电视台、广东电视台、凤凰网、中国网、中国新闻社、人
网、《南方都市报》、《南方日报》、《深圳特区报》、《深圳商报》、《广州日
》等近百家主流媒体聚焦本届颁奖盛典，在社会各界引起了广泛关注和强烈反应。

December 6th, 2016, world-famous International Space Design
ward—Idea-Tops held the 7th Awarding Ceremony in The Coli Hotel in
nggang District of Shenzhen. More than 1500 designers and well-known
trepreneurs from 35 countries and regions, including China, America,
K, France, Russia, Japan, Greece, Austria and so on, have participated.
ea-Tops 2016 has received 6020 entries from 35 countries and regions,
ch as China, America, Italy, Germany, Netherlands, Portugal, Belgium,
pan, Australia, Spain, Austria, Greece, Russia, Mexico, India, Romania,
ael, Brazil and so on. Top architects from 14 countries, including Britain,
stria, Israel, Egypt, Nigeria, Armenia, Serbia, Italy, France, Germany,

深圳市龙岗区委副书记、龙岗区人民政府区长戴斌致欢迎词

艾特奖国际学术委员会执行主席、哈佛大学建筑学博士、台湾亚洲大学副校长刘育东致辞

艾特奖组委会执行主席赵庆祥致辞

艾特奖国际评审代表——意大利共和国文化艺术总统勋章获得者 Michele mole致辞

深圳市人大常委会副主任蒋宇扬（右）、
深圳市政协常委、市政协提案委员会主任余立功（左）为获奖者颁奖

深圳市政协常委、科教卫委员会副主任马建文（右）为获奖者颁奖

深圳市科协党组书记、科协副主席张莉（右）、深圳市人民政府新闻办公室主任、"设计之都"推广办公室主任韩望喜（左）为获奖者颁奖

深圳市人民政府新闻办公室副主任、"设计之都"推广办公室副主任邱干（右）龙岗区政府文化产业发展办公室副主任闵玉辉（左）为获奖者颁奖

深圳市文体旅游局副局长钱强（左）为获奖者颁奖

深圳市人大常委会副主任、市科协主席蒋宇扬（右）、
龙岗区委常委、宣传部长胡庚祥（左）为获奖者颁奖

主办单位：艾特奖组委会 龙岗区人民政 导单位：深圳市委宣传 文体旅游 深圳市

办：深圳市东 传播有限公司、 坚合传

协办单位 莞市弘莹灯饰有 传单位

The 3rd G10 Designers Summit

第三届G10设计师峰会

12月6日，以"融汇·思变"为主题的2016第三届G10设计师峰会在深圳市龙岗区中海凯骊酒店隆重举行。汇聚30城力量、百位设计名家以"融汇·思变"的重要主题，再启百家争鸣。

G10设计师峰会由Idea-Tops艾特奖组委会于2014年首次倡导并发起，定位每年1届，盛邀100位不同地区、不同文化背景且极具代表性的设计师以圆桌小组讨论的形式，针对当下世界设计业的发展格局、现状、热门话题，从多层面、各角度进行深度研讨，并由小组组长做总结汇报。G10设计师峰会旨在为每一位参会设计师提供一个平等的发声机会、思想碰撞以及交流合作的平台，以此引领设计业的发展，推动世界设计产业的共同进步。

衣托前两届的成功举办，2016年第三届G10设计师峰会围绕着设计的发展核心，以"融汇·思变"的重要主题，艾特奖推广大使、PAL Design Group 创始人及首席设计师梁景华担任峰会主持人，并与来自中国大陆、港澳台两岸四地的百位设计名家，从建筑至室内设计，结合文化、市场、互联网、房地产等关系，以当下11个密切关联设计行业发展的深度话题展开研讨，剖析行业脉络，直击未来发展要领

On December 6th, The 3rd G10 Designers Summit was held at Coli Hotel in Longgang District, Shenzhen. With a theme of "Think Different", the summit gathered 100 famous designers from 30 cities to share their ideas.

G10 Designers Summit was advocated and initiated by Idea-Tops Organizing Committee in 2014. It is annually held during the Awarding Ceremony of International Space Design Awards—Idea-Tops, 100 representative designers from different regions and cultural backgrounds are invited to participate in the round-table discussions on industrial development, status and hot topics of the current global design industry. The summit aims to explore future trends of the industry, avoid blind spots and lead innovation on design. It provides an equal opportunity for every attendee to express ideas and to communicate and cooperate with each other, thus leading the industrial development and promoting the progress of the world's design industry.

Based on previous two summits, the third summit was focused on the development of design with the theme of "Think Different". Patrick Leung, spokesman of Idea-Tops, Founder and Chief Designer of PAL Design Consultants Ltd., was the host of the summit. 100 designers from Mainland China, Hong Kong, Macao and Taiwan got together to discuss 11 topics about architecture design and interior design, they combined culture, market, internet, real estate and other factors to analyze the industry context and emphasize the main points on future development.

Idea-Tops
International
Design Forum

艾特奖国际学术委员会

2016年12月6日，艾特奖国际学术委员会第二届学术会议在深圳召开。

艾特奖国际学术委员会是由艾特奖组委会携手全球22所顶尖学府包括哈佛大学、牛津大学、剑桥大学、东京大学、麻省理工学院、巴黎大学、罗马大学、清华大学、北京大学等一流院校的设计专家及教授于2015年共同发起成立。中国科学院院士郑时龄担任学术委员会主席，清华大学美术学院院长鲁晓波担任副主席，哈佛大学建筑设计博士、台湾亚洲大学副校长刘育东担任副主席兼执行主席。

此次经过艾特奖国际学术委员会批准，艾特奖国际学术委员会新增7位学术委员，丰富了国际学术委员会组织架构，为国际学术委员会注入更多智库资源。

至此，艾特奖国际学术委员会共有31位殿堂级学术委员，他们共同组成学术委员的强大阵容，为艾特奖建立了强大的学术支撑，发挥艾特奖在设计领域的专业影响力。未来将有更多来自全球在室内及建筑研究领域上独树一帜的行业领军人物成为艾特奖国际学术委员会成员，全方位、系统化地为国际学术委员会注入源源不竭的智慧资源，用学术的力量真正推动设计行业的革新与发展。

The 2nd Idea-Tops International Design Forum was held in Shenzhen in December 6th, 2016.

Idea-Tops International Design Forum was initiated by Idea-Tops Organizing Committee, well-known professors and academic leaders from 22 top universities in the world, including Harvard, Yale, Cambridge, Princeton, MIT, Tsinghua University and Peking University. Zheng Shiling (Academician of the Chinese Academy of Sciences) serves as Chairman of Idea-Tops International Design Forum, Lu Xiaobo (President of Academy of Arts & Design, Tsinghua University) serves as Vice Chairman, and Liu Yu-Tung (Doctor of Design at Harvard University, Vice President of Asia University, Taiwan) serves as Vice Chairman and Executive Chairman.

After the approval of Idea-Tops International Design Forum, 7 new academic members joined the forum, which enriches the organizing structure of the forum and adds more intellectual resources to it.

By now, Idea-Tops International Design Forum has 31 prominent academic members altogether. They form a strong academic member lineup, which builds a strong academic support for Idea-Tops award and increases the professional influence in design industry. More leading figures in global interior and architectural research fields will be academic members of the forum, and they will add endless intellectual resources comprehensively and systematically and use academic power to really promote the revolution and development of design industry.

The Idea-Tops International Design Forum 2016 Helped Chinese Design to Move Towards a New Journey

2016艾特奖国际大师论坛 助推中国设计走向新征程

2016年12月6日，作为本届艾特奖颁奖盛典的核心活动内容之一，国际大师论坛以"21世纪的建筑与城市"为主题全天呈现。吸引了国内外设计师、专家学者、相关机构负责人、行业媒体等近千名观众与会聆听。

依托艾特奖国际学术委员会的强大学术平台，2016艾特奖国际大师论坛盛邀了国外著名学府的专家教授发表了精彩演讲，同时参加2016年第15届威尼斯建筑双年展的14个国家著名建筑师也都重磅同台，带来了各自专业研究领域的主题演讲，众彩纷呈。

此外，2016艾特奖国际大师论坛围绕着"21世纪的建筑与城市"主题，分别举行了《21世纪的建筑与环境的关键方向》、《21世纪城市与建筑的冲突与矛盾》、《对话2016威尼斯建筑双年展》的三场巅峰对话，呈现了一场对21世纪的建筑与城市的顶级思考盛宴。

艾特奖作为一个强大的发展平台，吸引两岸四地乃至全世界顶尖的学府专家，优秀设计师代表、设计企业高管共同深究学术课题，它以中国为基础，作为世界设计建设的重要角色，助推中国设计走向新征程。

As one of the core activities during the Idea-Tops Awarding Ceremony 2016, the International Design Forum was held on Dec. 6th with a theme of "The Architecture and City of 21st Century". Nearly a thousand audiences attended the form, including domestic and overseas designers, experts, scholars, association leaders, media, etc.

Experts and professors from famous universities have been invited to make speeches in the Idea-Tops International Design Forum, and architects who exhibited at the 15th Venice Architecture Biennale shared a stage and contributed to a wonderful event.

Furthermore, under the theme of "The Architecture and City of 21st Century", three panel discussions on topics of "Key Direction of Architecture and Environment in the 21st Century", "Conflict and Contradiction of City and Architecture in the 21st Century" and "Experience 2016 Venice Architecture Biennale" were held during the forum. The forum stimulated deep thinking about the architecture and city of 21st century.

As a strong development platform, Idea-Tops attracts experts, excellent designers and design firm leaders from all over the world to discuss different topics. It's based in China and works as a constructive role for the world's design community, and it helps Chinese design to move towards a new journey.

16 THE SECOND IDEA-TOPS INTERNATIONAL
ESIGN FORUM

单位：艾特奖组委会　　　　　　协办单位：圣嘉佰利·东莞市弘莹灯饰有限公司　　　联合承办：深圳市东方辉煌文化传播有限公司
　　　深圳市龙岗区人民政府　　　支持单位：深圳市科源建设集团有限公司　　　　　　　　深圳康利置地有限公司
单位：深圳市委宣传部　　　　　　　　　　　费罗娜水泥砖、索菲亚磁砖
　　　深圳市文体旅游局　　　　　　　　　　菲尔顿瓷砖
　　　深圳市设计之都推广办公室
　　　深圳市科学技术协会

058
The
Best
Residential
Architecture Design Award
最佳住宅建筑设计奖

艾特奖

最佳住宅建筑设计奖
THE BEST RESIDENTIAL
ARCHITECTURE DESIGN AWARD

IDEA TOPS

INTERNATIONAL
SPACE DESIGN
AWARD

获奖者/Winner
Jean-Francois Milou
(新加坡/Singapore)
获奖项目/Winning Project
荷兰公园住宅/House at Holland Park

获奖项目/winning Project

荷兰公园住宅
House at Holland Park

设计说明/ Design Illustration

此项目由Jean-Francois Milou设计，他是studioMilou公司的创始人及总建筑师。项目前身是1930年代新加坡殖民时期保存下来的住宅，改造之后得以无缝延展成为当代多层居住和室外空间以及景观，满足了一个多代同堂的家庭需求。旧住宅的场地面积为2000平方米，位处一个广阔的三角地块上，是一座木框架砖砌体平房，被政府列为文化遗产。改造的主要挑战是将特殊形状的地块以及已存旧场地的限制，转化为一座联结多样化居住区域的住宅，同时传达一种广阔的、空间和谐的感觉。

充满景观的住宅
33荷兰公园是一座充满景观的住宅。走在住宅的每条过道上和公共区域里，景观随着步伐逐渐变化，空间与空间、主居住区和旧住宅自然衔接。植被点缀着大型玻璃窗，丰富的绿色植被与不同颜色的石头、墙和地板上光亮的缅甸柚木和谐地相互辉映。从新住宅的景观天台放眼望去，周围环绕着花园，树的顶端隐约透露出一种宏大的设计特质。

一条被丰富植被覆盖的弯道
项目克服了地块形状和已存旧住宅的限制，在新旧住宅之间、室内户外之间赋予了一种通透和流动的感觉。新结构的外墙是一个屏风般的外墙系统，包括一个上升屏幕，通往一条环绕整个场地的弯曲宽道，宽道及周边也种满了植被。住宅表面被丰富的植被覆盖，突出一种住宅蔓延进花园、花园栖息于住宅当中的感觉。密实的植被环抱着住宅的许多玻璃墙和通往二楼空间的塔楼。

一个私密的地方
从街上进入住宅，人们不会察觉到新住宅的宏大。新住宅被植被覆盖，每个区域经过精心设计，以避免与旧住宅产生突兀的对比。旧住宅区域为主待客区和厨房，人们从这里进入新住宅。旧住宅的出口是新住宅的入口。新住宅的外屏风墙旁是一个长方形水池，水和绿灰色的瓷砖在视觉上使得新旧住宅之间的联结变得柔和。

统一材质和颜色
白色天花板和墙——使旧住宅的室内十分简洁，整洁的家具——添加了一份朝向室外的开放性。贯穿新旧住宅的温暖的柚木地板构造了一种统一的联系，被一如既往地从各个方向朝向住宅的以及住宅内的植被点缀。室外选用的建材，包括灰色玻璃、混凝土柱和不锈钢，基于住宅整体与植被和光线的和谐关系以及住宅本身的通透感，被和谐地布置在植被当中。水池中水的反光和流动，柔软的灰绿色石材瓷砖，都增添了一份光线和生活的戏剧性，看上去同等地属于新旧住宅。

The design by Jean-Francois Milou, the founder and principal architect of studioMilou, begins with the conservation of a 1930s colonial residence in Singapore, and expands seamlessly from there into layers of contemporary living and outdoor spaces and landscaping, catering to a multi-generational family. The original house's site consisted of 2,000 square meters of broadly triangular-shaped land with a heritage-listed bungalow built of brick masonry with a wooden frame. Key challenges were to transform the unusual land shape and conservation limitations of the existing site into a home combining diverse living areas while conveying a sense of expansiveness and spatial harmony.

A house of views
33 Holland Park is a house of views. From each of the house's hallways and common areas are views which traverse and link one space to another, the main residence and the conservation house. Large glass windows frame the intense foliage at every opportunity, the still, rich greenery working with the palette of stone and polished Burmese teak of the walls and floors. From the landscaped roof of the new building, tree tops of surrounding gardens dominate, and discreetly contain the monumental nature of the design.

A winding path of abundant foliage
The project overcame the limits imposed by the land's shape and the existing conservation house with a design giving a sense of transparency and fluidity between the old and new buildings, between interior and exterior. The outer wall of the new structure is a paravent-like wall system, consisting of rising screens opening onto a wide path, winding around the site and lined with vegetation that ventures over the path and the neighbouring properties. A feeling that the house expands into the garden and that the garden inhabits the house is accentuated by the closeness of plants to the house's closed surfaces. Dense foliage caresses the many glass surfaces of the house, and towers to the second-floor spaces.

A private place
Upon entering the house from the street, the visitor is unaware of the monumental scale of the new house, covered as it is by vegetation and carefully designed proportions aimed at avoiding any stark comparison with the one-storey conservation house. The conservation house serves as the key reception area and kitchen, through which one passes to the new residence, the door of which is aligned with the exit of the former. A rectangular pool lines the outer paravent walling of the new house, the water and green-grey tiles softening the visual links between the conservation house and the new structures.
Unifying materials, colours
The simplicity of the conservation house's interior - white ceilings and walls, uncluttered furnishings - adds to a sense of openness towards the exterior. Warm teak floors throughout both buildings offer a sense of unity and complement the ever-present foliaging towards and into the house from all angles. The materials chosen for the exterior, including grey glass-reinforced concrete

columns and stainless steel, have been composed in such a way as to play with the colours of the vegetation and to accentuate the impression of an architecture which is transparent and in conversation with the trees and light. The reflections and movement of the pool's water, set off by the soft grey−green stone tiles, contribute to the play of light and life, and seem to belong equally to both houses.

获奖评语

采用了一种围合的
空间形式，营造出
一种场所感的氛围。
和自然之间形成一
种对话，创造了一
种新的建筑体验。

A kind of
enclosure
space form is
used to create
an atmosphere
of sense of
place, which
forms a
dialogue with
nature,
creating a new
architectural
experience.

NOMINATION FOR THE BEST RESIDENTIAL ARCHITECTURE DESIGN

最佳住宅建筑设计奖提名奖

Naoi Architecture & Design Office (日本/Japan)

获奖项目/Winning Project

辻堂住宅
Hut in Tsujido

设计说明/ Design Illustration

这个项目的基地是一片空置的、未被开发的土地，并与现有的郊区住宅毗邻。幸运的是，这个场地的周围被城市管理局作为绿色空间管理着，并将被保留下来。客户也希望在他们的日常生活中能保持这种自然的感觉，同时享受园艺时光和在屋外用餐的乐趣。于是，设计公司提出设计一个与风景相呼应的、简单的、像小屋一样的住宅。在屋内时，可以欣赏户外的美景，当站在屋外，你会赞叹它成为了自然的一部分。

第一层包括3个区域：起居室、餐饮区和生活区，每个区域都被裸露混凝土组成的地板空间的缝隙区分开来，使人体会到从外部空间到室内空间的连续性。第二层为卧室，明确界定为私人空间，如隐藏的阁楼，与第一层空间开放的氛围形成对比。

为呼应场地的地形条件，楼层被设计为跃层。餐饮区则为下沉式设计，这使得餐桌和露台在同一高度，从而与花园相互融合，在厨房，人们可以欣赏到花园的美丽景致。

房屋的开口处装有格栅窗，通过这个建筑结构，周围的自然景物象征性地通过格栅和框架被带入日常生活中。结构上的横梁开口使得白天拥有柔和的光线，为室内空间营造出开阔的氛围。

呼应"小屋"的设计理念，这座房子的主要结构是木材，而屋顶则设计成一个由垂直的钢支柱组成的简单桁架结构。为了保持倾斜的天花板下室内空间的开放性，屋架外露，但薄钢桁

架结构仍保持精致，与室内空间相协调。

房屋内每一个设计元素的积累和平衡，使住户享受到精致但简单、朴实的生活。即使在技术或价值观变化的时代，我们也希望这种架构，能实现大家想要感受自然的共同愿望，并且在很长的一段时间都能被欣赏和喜爱。

The site of this project was a vacant, untouched plot of land next to the existing suburban residence. Fortunately, the surrounding of the site is under the management of the city authority as a green space, and it will be kept as it is in the future--therefore the client hoped to maintain the feel of nature in their daily lives, while enjoying the time spending on gardening or having meals outside of the house. Then, the design company proposed a simple, hut-like residence corresponding to the scenery; from inside, the surrounding nature is to be enjoyed as a view; when staying outside, being the integral part of the nature is to be appreciated.

The 1st floor consists of three zones, including a living area, a dining and kitchen area, and a utility area, while each area is divided by a slit of 'doma' space – dirt floor space in exposed concrete finish–, allowing them to have the sense of continuity from the exterior space to the interior space. The bedroom is allocated on the 2nd floor, clearly distinguished as a private space to be hidden like an attic, compared to the open atmosphere of the spaces on the 1st floor.

The floor levels are designed to be a skipfloor, in corresponding to the topographic conditions of the site. The dining and kitchen area is sunken into the ground--this enables the dining table and the terrace to be at the same height, giving integrated feeling with the garden, while the garden and the scenery beyond the site may be viewed at the eye level from the kitchen.

The openings of the house are fitted with grid windows, so that the surrounding nature can be taken into the daily life as symbolic, filtered and framed view through this architectural gesture. The transom openings produced by the structural conditions allow to bring in soft indirect light throughout the daytime, providing openness to the interior space.

Responding to the concept of a "hut", the main structure of this house is produced in wood, while the roof is designed to be a simple truss structure with vertical struts in steel. In order to maintain the openness of the interior space with the sloped ceiling, the roof truss is exposed, yet the thin steel frames of the trusses maintain delicacy and make it possible to be harmonized with the interior space.

The accumulation and balance of each design element of the house enables the residents to enjoy sophisticated, yet simple and unpretentious daily lives. Even if technologies or the sense of value change in times, we hope this architecture, designed to fulfill the universal desire to feel nature, will be appreciated and loved for a long time.

NOMINATION FOR THE BEST RESIDENTIAL ARCHITECTURE DESIGN
最佳住宅建筑设计奖提名奖

张建（中国济南）
Zhang Jian (Jinan, China)

获奖项目/Winning Project

云生处
Where There are Clouds

设计说明/ Design Illustration

将南侧景观一侧围墙去掉，引入庭院。西侧墙体加高一直延伸至室内。而内部则保留了山区特有的砌筑方式，顶部采光则改变了山区房子北侧采光差的弊端。

In this project, the designer demolished the south wall to introduce a courtyard. The north wall has been increased and stretched itself to the interior. The inside spaces retained the special style of placing and masonry in the mountain. Usually the lighting condition of the north side of a mountain house is bad, but the current top light avoids such disadvantage.

NOMINATION FOR THE BEST RESIDENTIAL ARCHITECTURE DESIGN

最佳住宅建筑设计奖提名奖

Agraz Arquitectos SC.
(墨西哥/Mexico)

获奖项目/Winning Project

树之家
The Tree in the House

设计说明/ Design Illustration

这是一个特别的住宅，因为除了家人之外，里面还住着一棵树。房屋临近街道，表面看来位置不合理，但事实并非如此。房子的最佳方位是南北朝向，而这个住宅恰好是南北朝向，并与街道垂直，这一点也决定了该住宅应更加注重室内设计。设计师在建筑和街道中间设计了一块重要的过渡区域，可以让人在进入室内之前放慢步调。这一过渡区域体现的设计理念是，不管项目的面积有多小，设计师都有责任在城市中对空间进行创造。

住宅的朝向决定了其布局，大体上来说，主要方案是"L"形结构。面积较小的区域是入口和家庭活动室，较大的区域面向花园，设有厨房、餐厅、客厅和露台。而中间的主要区域清晰地划分出房屋的横纵界限，同时，营造出开阔的氛围。这一区域有着双层高的天花板，同时，高大的窗户旁栽了一棵郁郁葱葱的树，透过格架，树木与花园仿佛在隔空对话。格架由艺术家Adrian Guerrero设计，舞动的光线透过不同的格子照进室内，为房屋奠定了独特的基调。

光线滤过格子和树木，让人想起时光的流逝。建筑是空间的见证者，树木是时间的见证者。树木就像是房屋的伴侣，它伴随着家人一起生活，一起变老，分享着他们的记忆。这是一个独特的住宅，因为它不仅仅是居住者的家，同时也是一棵树的家。

This is a very special house, among other things because its main inhabitant, besides the family that lives here is a tree. Before we go any further it is important to observe that the emplacement of this house seems to be at odds with the street. Although seemingly so, it is quite the opposite. In this sense, since its longitudinal axis is oriented east-west (perpendicular to the street), and the most favorable orientation for the house is north-south, it was decided that the house accordingly so, should to look inwards rather than outwards. This means that it is not so much at odds with the street, but instead of looking directly at it, the project deals with the street through the creation of an important transition space between both the building and the street, which helps to change the pace while approaching the entrance. Underlying this exterior transitional space, lies the belief in the idea that each project, no matter how small, has a responsibility towards the creation of place in the city.

The decision to orient the house inwards determined its layout. In a broad sense, the main scheme is a very straightforward "L" figure. Its smaller section contains the entrance and family room, while the large part of the figure, that looks directly towards the garden, hosts the kitchen, dining room, living room and the terrace. In between them, lies the key space that articulates the house both horizontally and vertically, while at the same time, constitutes the house's main spacial feature. This is the only place in the entire building where we find a double height ceiling. It is so, because it contains a beautiful tree located next to the tallest window in the house. This is precisely the window that communicates, through its lattice, with the garden. Said lattice, designed by the artist Adrian Guerrero, combs the light and brakes it into a myriad of dancing sun rays that set the tone for the house's particular atmosphere.

Both the light filtering through the artwork and the tree, are meant to be daily reminders of time. As architecture is a witness of space, so the tree is a witness of time. The tree is this house's partner, it will accompany the family and grow old with them, it will share their lives and their memories. This is why it is a very special house, among other things because its main inhabitant, besides the family that lives here is a tree.

NOMINATION FOR THE BEST RESIDENTIAL ARCHITECTURE DESIGN
最佳住宅建筑设计奖提名奖

BCA Taller de Diseño
(墨西哥/Mexico)

获奖项目/Winning Project

迷雾森林之家
Casa Bosque de Niebla

设计说明/ Design Illustration

这个住宅位于墨西哥韦拉克鲁斯市，在设计公司接手项目时，一楼的墙壁和天花板已经在施工。这对设计师来说是一个挑战，他们必须在现有结构的基础上提出一个能满足客户需求的新方案。

为了提升空间的功能性，他们拆除了一部分墙壁。另外一个挑战是，客户希望这不仅仅是一个居住空间，它还应促进家庭成员之间的交流，并尽可能为日常生活创造更多的公共空间。

设计师将重点放在室内装饰上，整个项目满足了各个空间的需求，所有元素的运用都旨在展现每个家庭成员的性格特征。色彩与材料协调相融，例如石灰岩、墨西哥奶油色大理石与胡桃木的巧妙搭配形成鲜明对比，为空间增添了魅力。自然光线在室内逐渐减弱，营造了一种更加亲密的氛围。

For the project Casa Bosque de Niebla in Jalapa, Veracruz, there was another proposal already in process, some walls and ceilings for the first level. The first challenge was using what already had and to develop a new proposal according to the Client's needs and the interior project left in our hands.
The design firm made somemodifications

and had to demolish some of the existing walls to improve the functionality of the space. The second challenge was to develop a project according to the needs of the family who will live space who was asking to create an environment that would encourage communication and as many as possible meeting points for the daily coexistence of all. Focused in the interiors, designers developed an architectural project that takes the form from all the needs in each space and the elements selected to imprint the personality of each member of the family. The color palette harmonizes with materials such as limestone, Mexican cream marble and walnut wood that adds an attractive contrast. The natural light was faded traps to achieve a more intimate atmosphere.

082

The
Best
Public
Architecture Design Award

最佳公共建筑设计奖

艾特奖

最佳公共建筑设计奖
THE BEST PUBLIC ARCHITECTURE DESIGN AWARD

IDEA TOPS

INTERNATIONAL
SPACE DESIGN
AWARD

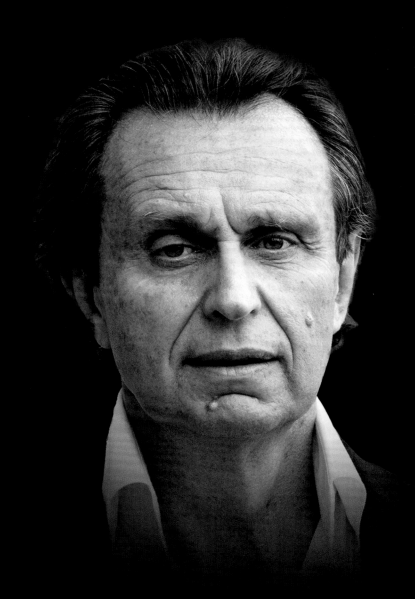

获奖者/Winner
Jean-Francois Milou
(新加坡/Singapore)

获奖项目/Winning Project
新加坡国家美术馆/National Gallery Singapore

获奖项目/winning Project

新加坡国家美术馆
National Gallery Singapore

设计说明/ Design Illustration

studioMilou事务所的新加坡国家美术馆设计将新加坡最重要的两座历史遗迹，前最高法院和前市政厅大楼，变成了东南亚最大的视觉艺术殿堂。客户要求将两座建筑合而为一并将其变成世界一流的艺术场馆，同时顾及其文物特点并尽量减少介入，尊重建筑的历史完整性。两座文物背负着巨大的情感和象征意义，其中包括第二次世界大战期间日军向盟军的投降以及1965年的新加坡宣布独立。

身为studioMilou首席设计师及创始人的法国建筑师Jean-François Milou，采用简洁的设计理念将整座场馆以一道简洁的屋顶线条把两座遗迹融合到一起。设计灵感来自手工精细编织的藤条，或传统编织布料。从街道上看，仅由一道淡淡的金色的屋顶柔和地披挂于两座历史遗迹，而由掐丝金属网罩组成的遮阳篷则微微地倾向两座遗迹之间的新入口。

美术馆内部通过控制自然采光以及颜色和材料的搭配来消除与历史特色之间的生硬对比，并在建筑风格多样的历史空间与新空间之间设置精微的变化，从而营造出一体化的氛围。为便于在如此规模的场馆中进行展览及各种项目，建筑师们打造了高效率现代化基础设施，与此同时却无损于历史遗迹的个性以及内部的典雅，所营造出的环境具有良好的通达性，让所有人都能享受并探寻馆内从老到新的各个层次上的历史与艺术。

可持续发展设计融入了整个美术馆的设计当中。通过透光的屋顶设计和窗户上使用的帘幕系统，过滤后的柔和光线渗透整个建筑物。在原有以及新添加的墙之间的双层设计大幅度地限制及减少了热量的增加，并鼓励自然光的渗透。位于空中花园内的简约水池有助于降低周围环境的温度，成为防火安全排烟口，同时让过滤的自然光进入下方的庭院空间。垂直的绿墙延续屋顶花园的长度创造出一个天然绿色空间，为建筑物的技术区域和机械冷却塔提供绿色屏幕。景观、垂直绿化和水景设计，不仅可以最大限度地减少太阳能热量进入建筑物，同时也增强声学设计。

studioMilou's design for the National Gallery Singapore transformed two of Singapore's most significant monuments, the former Supreme Court and City Hall, into the Southeast Asia's largest art visual arts institution. The client brief asked that the monuments be united into a world class gallery with minimal interventions, and respect for the historical integrity of the buildings. Enormous emotional and symbolic importance is attached to the monuments; the site of the Japanese surrender to the Allied Forces in 1945, and the site of Singapore's Declaration of Independence in 1965.

Jean-Francois Milou, the founder of studioMilou and principal architect, responded with simple design gestures: a filigree metal veil drapes over each building, a roofing structure linking the buildings from above as one institution. A design inspired, the architect's words, by finely woven rattan, or an ikat. Beneath the buildings, a sweeping concourse basement links them from below. From the street level, the monuments appear untouched but for a line of pale gold revealing the new roof, and the filigree awning that sweeps at the entrance.

Precedence is given to harmonious relations between the past and the present, to a seamless flow from the historic to the contemporary. The conservation work rigorously balanced the challenges of maintaining the original structures while accommodating the new functions of the gallery and offering radically new experiences of these monuments. The interiors share a limited colour and material palette, creating a sense of cohesion between the architecturally diverse historic spaces, the

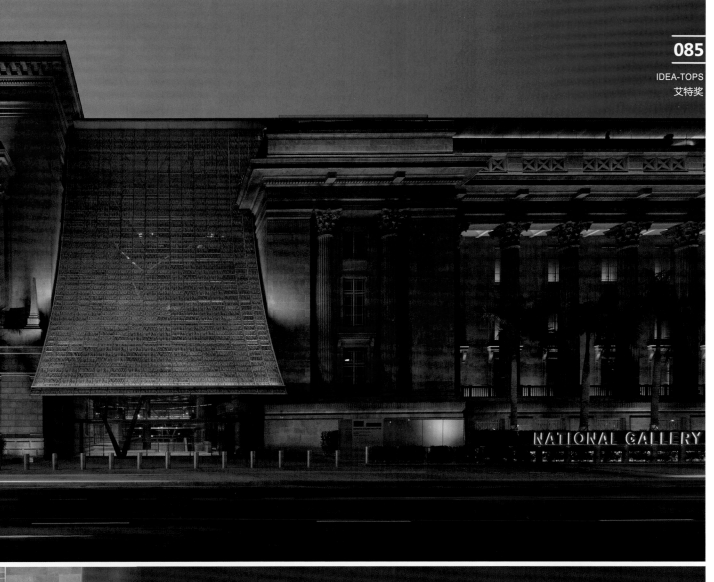

old and the new. Significant spaces within the former municipal and Supreme Court buildings have been restored - Justice Chambers, a rotunda reading room, former prison cells - beside new galleries. Complex technical facilities have been located in widened walls, within roofs and floors, with exacting care and stringent conservation requirements respected.

Sustainability is integrated throughout the Gallery. A system of softening and filtering the natural light through the roof-top veil and the screens used on the windows means natural light enters throughout the building, including in most of the exhibition spaces. Double-glazed facade design between the existing and new walls minimizes heat gain and encourages daylight penetration. Reflective pools placed within the rooftop garden provide a cooling environment for the roof-top, and serve functionally as fire-safety smoke vents, while allowing filtered natural light into the courtyard spaces below. Vertical green-walls create a continuous natural green canvas along the length of the roof-top garden, serving as a green-screen to the technical areas and mechanical cooling towers of the buildings. The landscape, vertical greenery and water features are designed to minimize solar heat gain into the building while enhancing acoustic design.

获奖评语

这是对城市的历史建筑的一个非常好的改造，巧妙的设计方法，提供了具有新加坡气候特点和地域风格的建筑。

This is a very good transformation of the city's historic buildings, with delicate design method, providing a building with Singapore's climate feature and regional style.

NOMINATION FOR THE BEST PUBLIC ARCHITECTURE DESIGN
最佳公共建筑设计奖提名奖

摄影：Mikko Ryhänen

Avanto Architects Ltd
(芬兰/Finland)

获奖项目/Winning Project

赫尔辛基Löyly桑拿房
Sauna Löyly in Helsinki

设计说明/ Design Illustration

桑拿文化
桑拿浴是芬兰文化和国家形象中不可或缺的一部分，仅有540万人口的芬兰却拥有着330万个桑拿房。然而伴随着拥有独立桑拿房公寓的逐渐普及，曾经随处可见的公共桑拿房正在急剧地减少。

环境
建筑的场地十分独特，距离城市中心不到两公里却同时有着如同偏远海岛一般的宽阔景观视野。将来这片顺着海岸延伸的场地将成为赫尔辛基公园的一部分，连接首都城市和大海。狭长形的布局顺着场地纵向伸展，以避免切断同样狭长的公园空间。低矮的建筑体量为未来住宅区预留出宽阔的海景视野。随性、多面化的设计让其完全融入了公园景观中，而不再是传统意义上的建筑。在经历了时间冲刷之后，木质建筑的色彩将慢慢淡去，那时它将如同耸立在海岸边的一块岩石。

建筑
建筑的概念非常简单：黑色的长方形盒子之内是温暖舒适的室内空间，而其之外，自由形态的木质外壳起伏伸展。由热处理的松木构成的结构不仅雕塑感十足，还十分实用。为室内的客人提供了隐私，同时，这些薄片并没有遮挡住室内看向大海的视线，而是像百叶窗一样阻挡了来自外面的视线。黑盒子和外壳之间的半覆盖空间成为了客人在桑拿浴间隙降温的场所。

外壳倾斜的部分之间形成了不同高度的观景、休憩场所。外层结构保护着内部空间不受寒冷的海岸气候的侵扰，阴影投落在室内空间巨大玻璃表面之上，降低了用于冷却建筑的能源消耗。外壳上阶梯状的台阶通往屋顶的楼梯和顶层观景平台，也将成为未来观看海上中心举行的运动赛事的大型室外观众席。大约4000块木板通过计算机控制的机器被准确地切割为不同的形状。巨大的木质平台延伸到海平面之上，汹涌的浪潮在平台之下拍打着海岸。

空间
建筑内部空间包含了两个部分：公共桑拿房和一间餐厅。桑拿和公共区域朝着海洋方向开放，可以看到城市中心和大海的有趣景观，略为昏暗的空间平静安宁。在一个整体的大空间中又包含了不同的区域，从一个区域到另一个区域可以感受到不同的氛围，餐厅宽敞、明亮，沐浴区则较为昏暗。

室内设计
餐厅和桑拿休息室的室内设计由Joanna Laajisto Creative Studio完成。设计师致力于用柔和的极简主义的设计方法，营造一个符合整体建筑风格的有气氛的餐厅。人们往往通过靠墙而坐来获得安全感和舒适感，而本次设计最大的挑战在于在一个两面临窗的开阔大厅里创造出一片私密的就座区域。设计师在酒吧区域建立了一个将空间一分为二的抬高的平台，而定制的沙发环绕幕墙而设，眺望着不远处的宽阔海景。
室内应用的主要材料为黑色混凝土、浅色的斯堪的纳维亚桦木、黑钢和毛料，所有的材料都耐用持久。芬兰新发明的可持续方法，将胶合板工业产生的废料经过压制、胶合和轻度热处理生成了桦木。极富创意的制造过程将废料变成美观的回收材料，生产出的材料有着好看的浅色调和很强的持久性。

桑拿房
这里三种不同的桑拿房都通过木材加热：持续加热的桑拿房，早晨在桑拿房开放前加热并整晚保温的一次性加热桑拿，以及在城市桑拿中十分罕见的传统烟熏桑拿。客人可以一次在Löyly经历各种不同的体验。桑拿房之间有一个带冷水池的spa区域和一个带壁炉的休息室，供客人在桑拿浴中途或之后使用。客人可以在海中游泳，甚至在冬天跳进在冰上凿出的洞内游泳。

可持续
建筑使用区域集中供暖和由水能和风能产生的电力，是芬兰首个、斯堪的纳维亚第二个获得FSC认证的建筑。森林管理委员会的认证证明了木材都是来自合法管制的森林。餐厅则供应有机食物和有节制地捕捉的鱼类。

Sauna Culture
Sauna bathing is an essential part of Finnish culture and national identity. There are only 5,4 million Finns but 3,3 million saunas. Public saunas used to be common in bigger cities but now that most new apartments have sauna of their own, public saunas have decreased dramatically in number.

Context
The site is unique. Being less than two kilometers away from the city centre, it is very central but at the same time the landscape is like in the outer archipelago. The plot is situated in a future coastal park that will be part of a broader "Helsinki park" connecting the capital city to the sea. The building was designed to be slim and elongated so as not to cut the narrow park strip. The volume is kept as low as possible so that it doesn't block views from the future residential blocks. Instead of building a conventional building, the sauna is developed into an easy-going, faceted construction that is more part of the park than a conventional building. When the wooden building turns gray, it will become more like a rock on the shoreline.

Architecture
The architectural idea is simple: there is a rectangular black box containing the warm spaces that is covered with a free form wooden "cloak". Instead of being mere decoration, the sculptural structure made of heat treated pine has several functions. It provides people with visual privacy. However, the lamellas don't limit the sea view from

inside it, rather they function like venetian blinds and blocking the views from outside. There are sheltered outside spaces between the warm mass and cloak to cool down in between sauna bathing. The cloak forms intimate terraces between its slopes that serve as a place to sit. The structure protects the building from the harsh coastal climate. It shades the interior spaces with big glass surfaces and helps to reduce the use of energy to cool the building. Moreover, the stepped cloak forms stairs to climb on to the roof and look out terraces on top of the building. The construction forms a big outdoor auditorium for the future marine sports centre's activities on the sea. There are around 4000 planks that were precisely cut to individual forms by a computer–controlled machine. The big wooden terrace is partly on top of the sea and you can hear the sound of the waves under your feet.

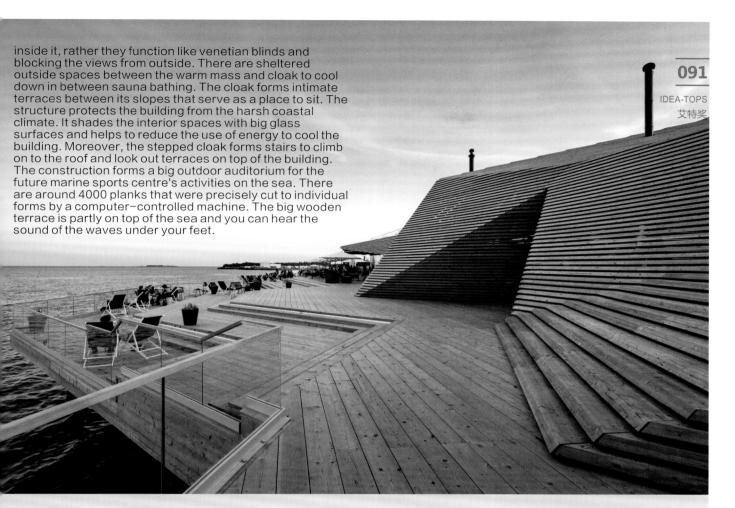

Spatiality
The building consists of two parts: public saunas and a restaurant. The saunas and public spaces open up to the sea, with interesting views to city center and even to the open sea. The atmosphere is calm and the spaces dimly lit. Different areas are conceived as spaces within a space. Interesting views open between closed spaces as you move from one area to the next. You enter in the restaurant that is a light and open space. From there a dimly lit sauna path leads to bathing area.

Interior
The interior architecture of the restaurant and the sauna lounge is by Joanna Laajisto Creative Studio. The objective of the design was to create an atmospheric restaurant which compliments the strong architecture of the building. The approach could be called soft minimalism. The challenge was to create intimate seating areas in the large hall like space with two walls of windows. People often feel most comfortable sitting their backs against the wall. The solution was to build a raised platform for the bar area which divides the space into two different areas. A wooden half wall anchors the long custom designed sofas which have a great view of the sea.
The main materials used in the interiors are black concrete, light Scandinavian birch wood, blackened steel and wool. All materials are durable and long lasting. The wood used is

pressed, glued and slightly heat treated birch, a new sustainable Finnish innovation made of left over materials of the plywood industry that normally is burned to produce energy. This is how waste is turned into a beautiful recycled material. Its manufacturing process produces a beautiful cool light color tone and heavy durability.

Saunas
There are three different saunas that are all heated with wood: a continuously heated sauna, a once heated sauna (that is heated in the morning before the sauna is open and stays warm all evening) and a traditional smoke sauna - a true rarity in an urban sauna. This is how you can experience all sorts of Finnish Löyly during a single visit.

Between the saunas there is spa area with cold water basin and a fire place room to relax in, between or after sauna bathing. You can swim in the sea and in the winter there is an "avanto", the hole in the ice for winter swimming - a popular hobby in Finland and our office name as well.

Sustainability
The building is heated with district heating and electricity is produced with water and wind power. The building is first FSC–certified building in Finland and second in Scandinavia. Forest Stewardship Council's certificate proves that wood material comes from responsibly managed forests. The restaurant serves organic food and sustainably caught fish.

NOMINATION FOR THE BEST PUBLIC ARCHITECTURE DESIGN

最佳公共建筑设计奖提名奖

Gpy Arquitectos
(西班牙/Spain)

获奖项目/Winning Project

拉古纳大学艺术学院
Faculty of Fine Arts, University of La Laguna

设计说明/ Design Illustration

西班牙拉古纳大学艺术学院的新校区位于校园边缘地带，毗邻的高速路、城市干道和校园道路带来熙熙攘攘的车流和人流。建筑师通过创造大量的开放性活动空间，使建筑完全融入至复杂的环境之中，实现了学校建筑与城市肌理的呼应和融合。建筑如同校园公共空间的延伸部分，蜿蜒的混凝土板在空中划出一道道圆滑的曲线，连接起两个园区，同时也界定了建筑的边界，围合出层次丰富、不受外界干扰的内部庭院。

架空的入口广场宽敞、开阔，俯瞰着两侧的繁茂内庭。或穿越广场到达对侧建筑，或沿两侧的走廊行走，或顺着蜿蜒的坡道下行，大量的人流在此聚集，再顺着环绕中庭的半开敞廊道，分散至建筑的各个角落。

走廊的另一侧是连续的教学空间。移动的分隔墙可以收拢至一侧，教室空间也将随之合并扩大，可根据需要改变尺寸，甚至贯通整个楼层。此外，大小不一的天井庭院、露天坡道、半开放的廊道以及入口广场平台都可兼作临时展览、教学和社交空间，灵活的建筑设计为空间的使用带来了更多可能性。

这栋全新艺术学院的颠覆性教学空间与校方创新实验型的教学方法相辅相成，成为了视觉艺术系学生最好的学习场所。

The new Faculty of Fine Arts is located in a heterogeneous area, adjacent to the island highway and on the periphery of the University Campus. The main challenge was to create a link between the new faculty building and its surroundings by working with the open public spaces and to increase the synergies between the academic complex and its urban context. The new building presents itself as an extension of the Campus's public space, while creating an autonomous interior landscape of its own. A skin of suspended concrete slats adopts a curved shape which develops on the different levels protecting and wrapping the open space of the building.

Campus circulation is collected and guided by a public plaza that extends through the building's main entrance and is transformed into a spacious terrace overlooking the inner courtyard. From the main entrance, circulation is continuous, following half-open, undulating corridors.

The teaching areas are distributed along a continuous band accompanying the open corridors and dispose of mobile dividing walls that allow for creating classrooms of different sizes or even opening up the whole floor, depending on the needs. Adding to this flexibility in use are multiple spaces like the patio-gardens and open ramps, the covered galleries and the entrance terrace, conceived as open exhibition and teaching areas and places for social exchange.

The designers like to see the new Faculty of Fine Arts as a building that offers ground-breaking, innovative spaces for experimental and creative education of future students of visual arts.

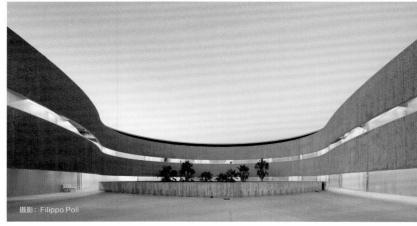

摄影：Filippo Poli

摄影：Jos é Ram ó n Oller

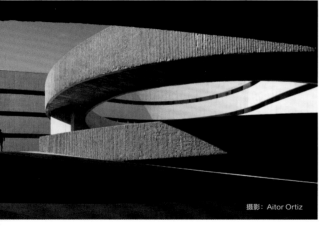

摄影：Aitor Ortiz

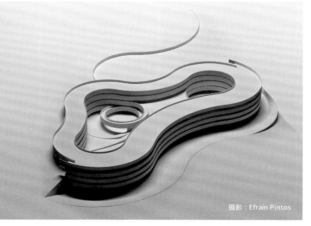

摄影：Efrain Pintos

摄影：Filippo Poli

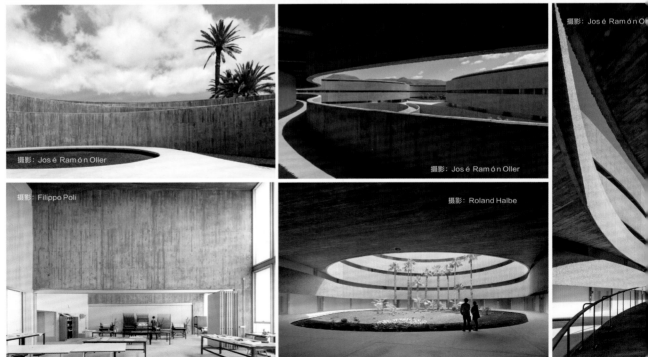

摄影：José Ramón Oller

摄影：José Ramón Oller

摄影：José Ramón Ol

摄影：Filippo Poli

摄影：Roland Halbe

摄影：Roland Halbe

摄影：Filippo Poli

NOMINATION FOR THE BEST PUBLIC ARCHITECTURE DESIGN

最佳公共建筑设计奖提名奖

McBride Charles Ryan

(澳大利亚/Australia)

获奖项目/Winning Project

艾文豪文法学校高年级科学中心
Ivanhoe Grammar Senior Years & Science Centre

设计说明/ Design Illustration

艾文豪文法学校是一所创立于1920年的男女混合学校。校区面积非常宽广，附近被一大片赤桉所环绕，尽显乡村田园风情。尽管濒临郊区，校园仍保留了当地的景观特色。在校园中央区域，原先的建筑以一种类似城市广场的结构正式地排列着。具有不同建筑风格的元素逐渐加入到校园内，为这个正式的中心增添了轻松的氛围。

设计公司的意图是为学校打造一个全新的高年级科学中心，其中包括为高年级教师提供一系列综合学习区，以及为年龄较小的学生设立一个科学中心。

基于学校最初的总体规划，建筑采用了圆形结构，这种结构极受建筑师的青睐，它就像是人类在景观中留下的印记。然而，设计公司并没有完全采用圆形或半圆结构，而是在其中加入了一个多角度的几何结构。这一几何结构划分出中央庭院、采光井和镶有马赛克的学习区，切断了建筑的圆形结构，与之形成反差，让入口的设计变得非常醒目，并与外部空间明显地区别开来。内部是一个充满活力的、富有表现力的、丰富多彩的综合区，而外部则是一个柔和的、单一的圆形区域。

设计公司重点考虑了学习区的布局，最终该区域呈现出非常明显的特征，即具有互联性、多功能性、灵活性和适应性。建筑内各个区域都设置了工作区，较低的楼层大部分是多功能教学空间，较高的楼层则以科学空间为主，此外还有一些非正式区域供学生休闲。

所有楼层由钢筋混凝土、环梁和钢柱构成，楼顶采用了钢架，外立面运用了Colorbond Longline钢板，内部运用了Vitrapanel建筑板。外部悬挂的"鳍"是昆士兰胶合板桉板，窗户安装了具有隔热效果的双层玻璃。另外，各个学习空间设有高级隔音玻璃门进行区隔。

为了让教育更加全面，设计师在现代教学方法中融入了对比强烈的建筑语言。圆形代表了典型的遵规守矩、严谨态度和知识的确定性；而内部区域中，色彩丰富、形态各异的马赛克折射出不同的光线，代表了现代生活和科学认知的不确定性和复杂性，以及在前进过程中拥有好奇心和想象力的重要性。

Ivanhoe Grammar School is a co-educational school established in Ivanhoe in 1920. The Plenty Campus of Ivanhoe Grammar School was, until quite recently, surrounded by a rural setting punctuated by magnificent red river gums. Despite the encroaching suburbs, the school has retained a character where the native landscape flavour dominates the campus. At the heart of the campus, the original buildings are set out in a formal arrangement based on the metaphor of the Town Square. Subsequent additions to the campus, executed in a variety of architectural styles, have a looser relationship with this formal centre.

The design firm's brief was for a new Science and Senior Years Centre. The brief included a variety of general learning areas, provision for the senior year teachers and a science centre which was to be also used by the younger students in the school.

The circular shaped plan form was adopted for the building; this shape had an appropriate civic quality which seemed to build upon the schools original master plan. The circular plan is an alluring one for architects, clearly it is a definitive human mark upon the landscape. However, rather than adopting a circular or radial pattern dictated by the shape of the plan, the firm chose to overlay an angular geometry. This geometry was used to define the central courtyards, the light wells and a mosaic of learning spaces. This geometry contrasts and disrupts the building's circular motif, highlighting key entry points and providing a distinction between the outer world (singular, civic, circular, executed in a muted landscape palate) and the inner world (complex, dynamic, expressive & colourful).

Great consideration was given to the configuration of these learning spaces. Some of the key characteristics were transparency into and between spaces, a variety of spatial type, interconnectivity, multiple-use, flexibility and adaptability of the learning spaces. Staff work stations are peppered throughout the facility; the lower levels are generally multiple-use teaching spaces, the upper levels predominately science focussed. Provision for outdoor informal gathering of students is provided throughout the complex.

The building floors are reinforced concrete with band beams supported by concrete and steel columns. The roof is steel framed. The outer cladding is Colorbond Longline, the inner cladding Vitrapanel. The exterior shading fins are glue laminated Queensland spotted gum. Windows are aluminium, double glazed & thermally broken. High quality acoustic glazed doors are used between learning spaces.

The contrast so evident in this building's language encapsulates the contemporary pedagogical approaches for a well-rounded education. The circular form is a classical representation of order, rigour, and the certainty of knowledge - the buildings inner world, with its complex mosaic of spaces, amplified by pattern, colour and multiple reflections, represents the uncertainty and complexity of modern life and scientific understanding, and the necessity of the qualities of wonder and imagination to advance and see us through.

NOMINATION FOR THE BEST PUBLIC ARCHITECTURE DESIGN
最佳公共建筑设计奖提名奖

王凡、张应鹏（中国苏州）
Wang Fan, Zhang Yingpeng (Suzhou, China)

获奖项目/Winning Project

浙江湖州梁希纪念馆
Liang Xi Memorial Hall in Huzhou, Zhejiang

设计说明/ Design Illustration

浙江湖州的群山中、幽谷里，国家首任林垦部（林业部）部长，杰出林学家、教育家和社会活动家梁希先生的纪念馆坐落于此。"纪念"对我们此次设计的主创团队而言，并不是高高在上的仰望及设计化的纪念结局，我们想给予大众的是一种开放式的纪念体验，一种多样的记忆留存。纪念馆和整个梁希森林公园的自然环境相融合，依山借势：建筑空间与山体相融合，建筑人行路径与游园路径相融合。开放的纪念线索设置：自由的参观路径，拼贴化的空间及故事的组合，演绎出了每个参观者各自独有的体验。

梁希先生描绘的"无山不绿、有水皆清、四时花香、万壑鸟鸣"的美好境界，也被观者以自己独有的情感体验而记忆传承。主创设计团队一体化完成了纪念馆建筑、室内、景观及展陈的设计，尽自己最大的努力向梁希先生致敬！（该项目其他设计参与者：钱弘毅、肖蓉婷）

Liang Xi Memorial Hall is located among the hills and valleys of Huzhou, Zhejiang. Liang Xi, the Republic's first minister of Department of Forest, was an outstanding forester, educator and social activist. For the design team of this project, "memorial" is not about the superior feeling or a designed result, but an experience that is open for the public,

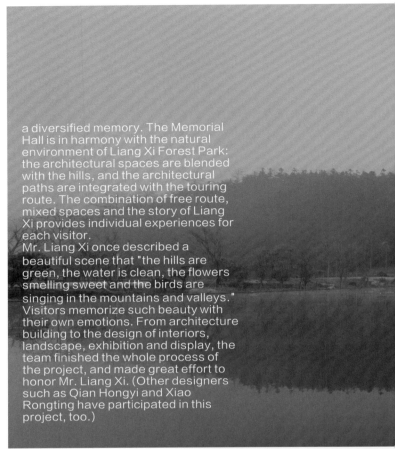

a diversified memory. The Memorial Hall is in harmony with the natural environment of Liang Xi Forest Park: the architectural spaces are blended with the hills, and the architectural paths are integrated with the touring route. The combination of free route, mixed spaces and the story of Liang Xi provides individual experiences for each visitor.

Mr. Liang Xi once described a beautiful scene that "the hills are green, the water is clean, the flowers smelling sweet and the birds are singing in the mountains and valleys." Visitors memorize such beauty with their own emotions. From architecture building to the design of interiors, landscape, exhibition and display, the team finished the whole process of the project, and made great effort to honor Mr. Liang Xi. (Other designers such as Qian Hongyi and Xiao Rongting have participated in this project, too.)

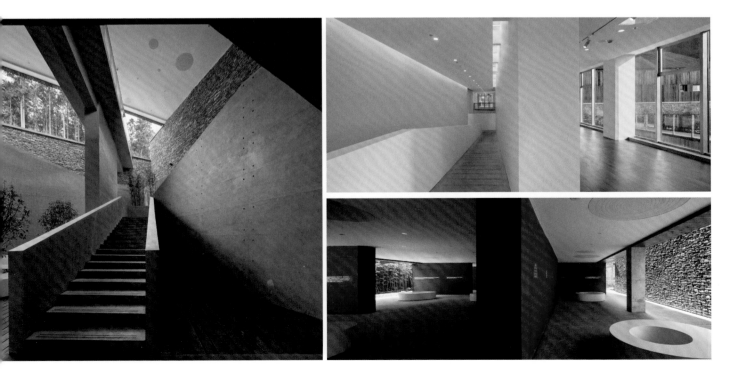

106

C

The Best
Green
Architecture
Design Award
最佳绿色建筑设计奖

艾特奖

最佳绿色建筑设计奖
THE BEST GREEN ARCHITECTURE
DESIGN AWARD

IDEA TOPS

INTERNATIONAL
SPACE DESIGN
AWARD

获奖者/Winner

Studio Marco Vermeulen
(荷兰/Netherlands)

获奖项目/Winning Project

Biesbosch博物馆/Biesbosch Museum Island

获奖项目/winning Project

Biesbosch博物馆
Biesbosch Museum Island

设计说明/ Design Illustration

这是一个重建项目，新加入了一个餐厅和一个临时的当代艺术展厅，使原来的博物馆完全向周围环境开放。

建设

为了避免一些不必要的材料浪费和资源消耗，设计团队保留了原博物馆展馆的六边形结构，在建筑的西南侧增加了1000平方米的新体块。扩建部分以大开窗为特点，此部分容纳了可以欣赏到邻近景观的有机餐厅以及举办临时展览的画廊。同时，已有的建筑用于常设展览，容纳了图书馆、多功能剧院、入口区和博物馆商店。

博物馆的新旧部分被土方包围，上面覆盖着草坪和香草植物。设计旨在增加生态价值，体量如雕塑般屹立于此，被看作是大地艺术，从景观中脱颖而出。屋顶上的弯折区域形成了山间小路和瞭望台。

能源

设计的整体性将能量消耗降到最低，装有耐热玻璃的玻璃窗取消了百叶窗的使用。同时，西北侧的大地艺术土方和绿色屋顶起到了额外的保温隔热作用。在寒冷的天气里，生物质炉可维持建筑的温度；而在温暖的月份里，来自河流的水源通过相通的管道流动，以此给博物馆降温。

柳木过滤器

通过场地上的柳树过滤净化生活污水。植被吸收了废水中的富营养化的氮和磷酸盐，有助于柳树的生长。然后净化后的水被排入邻近的湿地区，流入河流中。一旦柳树被锯断并晒干，木材可被用作生物质炉的燃料。

淡水潮汐公园

这个博物馆完工后将变成一个淡水潮汐公园，通过新挖的河道引入河水。博物馆的缓坡可以造成潮汐和水位的季节性变化。地形条件还形成丰富多样的动植物群，同时，随着水位的变化，蜿蜒的小路在外观上也不断改变。

The existing museum building is completely transformed and expanded with a new wing that opens the building to the beautiful environment. The new wing accommodates the restaurant and a temporary exhibition space for contemporary art.

Museum Building

To avoid unnecessary wasting of material and energy, the structure of the hexagonal pavilions of the original Biesbosch Museum is maintained. On the southwest side the building a new wing of 1000m2 is added. The expansion of 1000 sqm comes with lots of glass and thereby opening the building to the part of the island that is activated as a museum garden. The expansion accommodates an organic restaurant with views of the adjacent water and landscape. The other part is used as

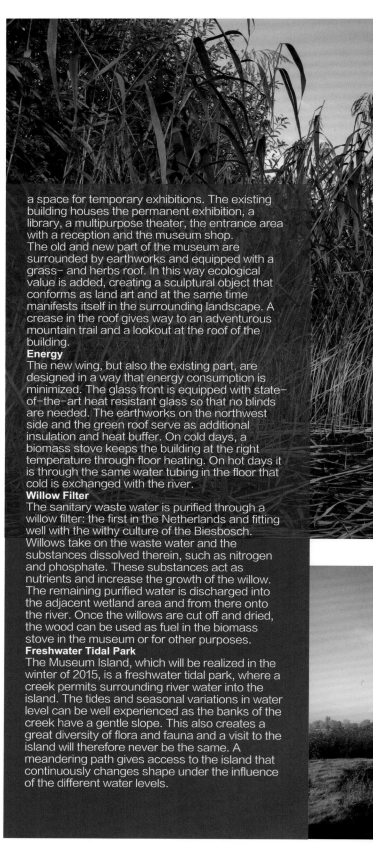

a space for temporary exhibitions. The existing building houses the permanent exhibition, a library, a multipurpose theater, the entrance area with a reception and the museum shop.

The old and new part of the museum are surrounded by earthworks and equipped with a grass- and herbs roof. In this way ecological value is added, creating a sculptural object that conforms as land art and at the same time manifests itself in the surrounding landscape. A crease in the roof gives way to an adventurous mountain trail and a lookout at the roof of the building.

Energy

The new wing, but also the existing part, are designed in a way that energy consumption is minimized. The glass front is equipped with state-of-the-art heat resistant glass so that no blinds are needed. The earthworks on the northwest side and the green roof serve as additional insulation and heat buffer. On cold days, a biomass stove keeps the building at the right temperature through floor heating. On hot days it is through the same water tubing in the floor that cold is exchanged with the river.

Willow Filter

The sanitary waste water is purified through a willow filter: the first in the Netherlands and fitting well with the withy culture of the Biesbosch. Willows take on the waste water and the substances dissolved therein, such as nitrogen and phosphate. These substances act as nutrients and increase the growth of the willow. The remaining purified water is discharged into the adjacent wetland area and from there onto the river. Once the willows are cut off and dried, the wood can be used as fuel in the biomass stove in the museum or for other purposes.

Freshwater Tidal Park

The Museum Island, which will be realized in the winter of 2015, is a freshwater tidal park, where a creek permits surrounding river water into the island. The tides and seasonal variations in water level can be well experienced as the banks of the creek have a gentle slope. This also creates a great diversity of flora and fauna and a visit to the island will therefore never be the same. A meandering path gives access to the island that continuously changes shape under the influence of the different water levels.

获奖评语

这个建筑对绿色建筑类别来说非常完美，它在环境、景观和建筑本身之间制造了一种强烈的联结。这很有趣，这是建筑的新边界。

This project is perfect for the green architecture category. It produces a strong relationship between the environment, the landscape and the new architecture. And this is really interesting. This is the new frontier of architecture.

NOMINATION FOR THE BEST GREEN ARCHITECTURE DESIGN

最佳绿色建筑设计奖提名奖

Atelier Tekuto (日本/Japan)

获奖项目/Winning Project

R · Torso · C

设计说明/ Design Illustration

首座使用SHIRASU（白砂）混凝土的建筑

这个住宅位于东京中心部，用地面积66平方米。客户是一对化学家夫妇，兴趣是建筑与艺术。他们最初是希望建造"被混凝土所包围的空间，并且是使用至今没有过的环保型混凝土。"

在东京的高密度住宅区，比如项目所在的区域，享受大的空间是一种奢侈，更不用说广大的大自然。因此我们能接触的自然只能是天空了。我们一直在反复研究住宅与天空的关系。因为建筑面积只有30平方米，对着天空的大开口将成为一种有效手段，使室内空间带给人的感觉变得广阔。最后我们切开的这个大开口不仅使室内的人感受到更广阔的空间，也使路上的行人视线可穿过切口看到更广阔的天空。

为了确保舒适的室内环境，我们与环境工程师山田博之进行了合作，设置了"热循环系统"。该系统利用了四层打开的楼梯井使每层楼空气流通。在冬季，使用恒温箱和可逆的风扇让房子上部温暖的空气通过通风管道自动传送到地下室。在最冷的时候，在通风道加入1个电散热器，增加

所需要的额外热量。在夏天，通风道的冷空气发送到三楼，空气向下流入通风道内再次冷却整个房子。

建筑本身是由100%的可回收混凝土（SHIRASU混凝土）构成，有助于构建一个低碳和资源可回收的社会。
这个项目里，我们提供了100%可回收环境型SHIRASU混凝土，SHIRASU混凝土是利用沉睡在日本南九州的火山灰SHIRASU所制成。这种混凝土的优点是它的强度和耐久性，因为SHIRASU的火山灰反应与SHIRASU的特殊密度，它还包含一些微封闭细胞，有利于湿度和气味的调节。开发和利用SHIRASU混凝土对于可开采SHIRASU的地区来说是一笔很大的财富。

而且由于它是自密实混凝土，它降低了施工现场的噪音、振动和能源。R · TORSO · C是世界上首座使用了环保混凝土，并结合了热循环系统的建筑。

通过材料思考社会
社会背景
在20世纪，混凝土是建筑行业的驱动力，它一直与经济同步增长。

据预计，作为混凝土主要成分的水泥产量将持续上升，2050年将达到20世纪用量的一倍之多。此外，水泥是通过煅烧加热与其他材料的天然石灰岩生产的，将释放出大量的热能与所产生的二氧化碳。混凝土的骨料是砂石和砂粒，两者都是自然资源。而近代制造业开始使用来自海洋的砂石来获取而代之。这样的结果不仅使混凝土中的钢筋锈蚀，而且还包含许多氯化物来源于海水，使混凝土的质量普遍下降。此外，海滩砂的提取是被许多地方政府监管或禁止的，因为它对海洋环境造成恶劣的影响。

因此，寻找可以替代利用的资源来代替砂石资源是我们的课题。还要改善混凝土的回收过程，研发出更优良耐用、多功能的混凝土。

这是为什么我们会想到要开发SHIRASU混凝土。

环保材料"SHIRASU"具体的特点：
1.为保护自然资源做出贡献
日益稀缺的资源，混凝土中砂的60%~70%可使用"SHIRASU"来取代。

2.环保
该混凝土在拆除时的水泥原料可被完全回收利用。
减少CO_2的排放（关于水泥）和在煅烧时的能量排放。

3.硬度和耐久性
因为"SHIRASU"的生成经过了很长的时间，并经过火山灰反应，有很强的硬度与耐久性，并有抗硫酸与抗盐性。
此外，关于密度，"SHIRASU"的精细粒度可以保护混凝土中的中和作用。

4.表面光滑，可控制湿度的品质
其光滑的纹理可精加工成暴露在外的材质。
"SHIRASU"也包含微封闭细胞，可以使混凝土具有调节湿度与除臭的品质。

First Architecture built using "SHIRASU" Concrete

The house is located in the center of Tokyo, on a site merely 66 square meters in area. The clients are a married couple both working in the field of chemistry, sharing a passion for architecture and art. "We want to see exposed concrete finish inside and out. A distinctive piece of architecture that is at the same time environmentally conscious", were the initial requests of the clients.

In high density residential area in Tokyo, such as this site, you seldom have the luxury of large open spaces, let alone vast nature. The only direction that we can turn for those factors is the "Sky". We have been repeatedly studying this approach to the sky as an element of nature in previous projects and one way of establishing it is by pruning away the corner of a rectangular volume at an angle. Because the building footprint is only 30 square meters, large openings facing the sky becomes an effective means to incorporate the feeling of vastness onto the compact internal space. The chamfered corners, the result of which are large openings, not only provide the interior space with a view to the sky, but also allow for a more open view to the sky for the people on the street.

To secure a comfortable interior climate, a "thermal circulation system" was incorporated with the help of environmental engineer Hiroyuki Yamada. This system makes use of the stacking effect in the 4 story stair well that is opened to the living space on each floor. In the winter, the warm air that accumulates in the upper part of the house is transferred to the pit below the basement floor automatically through a ventilation duct and reversible fan connected to a thermostat. An electrical radiator is added in the pit for the extra heat needed in the coldest season. In the summer, the chilled air from the pit is sent up to the 3rd floor to flow down into the pit again cooling the entire house.
The building itself is made from **100% recyclable concrete, what contributes to a society of low carbon and recycling of resources.**
For this project we developed a 100% recyclable concrete which, instead of sand, contains SHIRASU. "SHIRASU" is a general term for the deposit of pyroclastic flow of volcanic ash found in south of Kyushu. The advantage of this concrete is its strength and durability that increases over a long period of time because of the pozzolanic reaction of SHIRASU. Also its density, which comes from the fine granularity of SHIRASU, protects the concrete from neutralization. SHIRASU also contains micro closed-cells which gives the concrete humidity control and deodorizing qualities. This development and use of SHIRASU concrete can be a huge asset to those areas where SHIRASU can be excavated.
As it's being self-consolidating concrete, it reduces the noise, vibration and energy at the construction site as well. The world's first architecture, R torso C was built using environmentally friendly concrete, as well as incorporating a sectional thermal circulation system.

Thinking about the environment through materials: Social Background
In the 20th century, the production of concrete which was the driving force for the building industry, it has been making constant increase in synchronicity with economic growth.
It is expected that the production of cement, being the main ingredient of concrete, will continue to rise. It is assumed that the consumption will be doubled by 2050 compared to the turn of the century. Additionally, cement is produced by heating natural limestone with other materials through a process known as calcination, a lot of heat energy is required that results in vast CO_2 emission. The aggregate for concrete are gravel and sand. Both are natural resources. Up until the economic boom, these materials were extracted from dry river beds or mountains resulting in drainage of good resources. Instead manufactures started to use sand from the seas. This results not only in the rusting of the reinforcement bars in concrete, but also contains many chlorides that originate from the sea water, which causes a general deterioration of quality of the concrete. Moreover, the extraction of beach sand is being regulated or banned by many local governments, because of the bad effects it has on the marine environment.
It is a matter of utmost urgency within the concrete

industry, clearly an industry that is indispensable to contemporary architecture, to search for alternative resource that could be utilized instead of sand and gravel. Also to change the formation of the recycling process of concrete, develop a more durable and multifunctional concrete.

This is how we came up with the idea for the practical use of volcanic "SHIRASU" as a fine aggregate.

Characteristics of Environmentally Conscious "SHIRASU" Concrete:

1. Contribution to the preservation of natural resources
60% to 70% of sand in concrete, which is becoming a scarce resource, can be replaced by "SHIRASU".

2. Environmentally friendly
The concrete becomes completely recyclable as a cement raw material at the time of demolition.
Calcination energy [of the cement] and CO_2 emission can be reduced.

3. Strength and durability
Strength and durability that increases to grow over a long period of time because of the pozzolanic reaction of "SHIRASU", resistant to sulfuric acid and salt-damage.
Also its density, which comes from the fine granularity of "SHIRASU", protects the concrete from neutralization.

4. Smooth surface, humidity controlling qualities
Smooth texture allows to have exposed concrete as fine finish.
"Shirasu" also contains micro closed-cells which give the concrete humidity control and deodorizing qualities.

NOMINATION FOR THE BEST GREEN ARCHITECTURE DESIGN

最佳绿色建筑设计奖提名奖

UNStudio & DP Architects

(荷兰 & 新加坡/Netherlands & Singapore)

获奖项目/Winning Project

新加坡科技设计大学
Singapore University of Technology and Design

设计说明/ Design Illustration

由UNStudio及DP Architects共同设计的新加坡科技与设计大学反映对现今教育机构面对不断改变的需求的深刻理解。因此，UNStudio便以连接度、协调性、共同创造、创新及社交性作为新校园的设计核心要素。

新加坡科技与设计大学是当地著名大学之一，提供了4个主要学科：建筑与可持续设计、工业设计、工程设计及信息科技与设计。而新校园能汇聚人才、想法及创意，使大学成为科技创新及经济发展的推动者。

以该校创意冒险精神促进跨学科的接口，新校园的设计直接反映了大学的课程。校园设计透过学生、教职员及教授之间的非线性联结，提供了机会去采纳创新与创意。

根据总体规划，新校园的设计包含了两个重要元素：生活与学习。这两个重要元素形成设计的中心点，把校园的每一个角落都联结起来。在这中心的接合点，UNStudio 提出了可举办展览、活动及增加互动性的灵活空间。于是，中心仿如校园的智能心脏，直接联结观众席的主要架构支柱、国际设计中心以及大学图书馆。

新加坡科技与设计大学是可持续性的建筑物，包括各种设计策略以减低新加坡热带气候所造成的影响。除了冷却技术、有盖行人道、百叶幕墙、开放空间、室内日光及应付滂沱大雨的保护，广泛的取向及自然通风原理的风向研究也被全面地应用于校园设计当中。该设计为SUTD校园获得了最高绿色建筑标志评级（白金）在新加坡使用。

Designed by UNStudio and DP Architects, the academic campus for the Singapore University of Technology and Design reflects an in-depth understanding of the changing requirements of learning institutions today. Connectivity, collaboration, co-creation, innovation and sociality are at the basis of UNStudio's design thinking on New Campuses.

The SUTD offers four key academic pillars: Architecture and Sustainable Design, Engineering Product Development, Engineering Systems and Design and Information Systems Technology and Design. The SUTD is a driver of technological innovation and economic growth, with the new campus bringing together people, ideas and innovation.

The new academic campus directly reflects SUTD's curriculum, using the creative enterprise of the school to facilitate a cross-disciplinary interface. The design offers an opportunity to embrace innovation and creativity through a non-linear connective relationship between students, faculty and professionals.

Following the master plan, the academic campus is designed through two main axes: the living and learning spines, which overlap to create a central point and bind together all corners of the campus. At the centre of this node UNStudio proposed a flexible space for exhibitions, events and interaction: this Campus Centre forms the intellectual heart of the campus and directly links the main programmatic anchors of the Auditorium, the International Design Centre and the University Library.

The SUTD is a highly sustainable building, incorporating numerous passive design strategies designed to counteract the conditions caused by Singapore's tropical climate. Based on extensive orientation and wind studies natural ventilation principles are applied throughout the design, alongside cooling techniques, covered walkways, louvred facade shading, open voids, extensive daylight to the interiors and protection from heavy rain showers. The design for the SUTD campus achieved the highest Green Mark rating (Platinum) available in Singapore.

NOMINATION FOR THE BEST GREEN ARCHITECTURE DESIGN

最佳绿色建筑设计奖提名奖

竹工凡木设计研究室（中国台湾）
CHU-studio (Taiwan, China)

获奖项目/Winning Project

日安崎聚
Flowing Space of Natural Environment and Concrete

设计说明/ Design Illustration

渗入透出的微型聚落

在地狭人稠的当代都市，城市规划者与建筑师几乎时时刻刻思索着：如何争取更多的居住空间？这个议题在20世纪现代主义提出以Domino系统为基础的钢筋混凝土高层建筑后似乎得到了暂时的纾困，而我们的城市也随之变成国际样式（International style）下的水泥丛林，就在容积、建蔽、公设比的精打细算中，大型高密度集合式的住宅楼群运而生，成为绝大多数已开发都市的典型地景。而人们的生活与思维，似乎也被迫缩身于与邻人划清界线的小框框里。偶尔人们会奢侈地想念起，过去农村里开阔宽敞的合院空间，晒着稻谷菜干的大院里，孩子们在阳光下嬉戏笑闹，大人闲坐一旁乘凉聊天，泡茶下棋。这番景象，难道真的只能是过去式了吗？

水泥丛林中的现代合院聚落

基地是位于新竹县宝山乡一处由自然环境所环抱的郊区，与新竹科学园区相距甚近，不仅交通方便，竹林绿带所围合出的隧道也成为最好进入此基地的入口意象。本案是旧建筑改造成新形态的聚落小区，我们希望将传统集合式住宅小区与合院结合，使自然与生活两者皆能兼具，创造出更丰富多元的生活环境。

游移的缝隙 共享经济的生活理念

本案是个微型聚落的计划，我们思索现代小区经营理念与传统合院聚落精神融合的可能性。现代集合式小区的特色是一个萝卜一个坑，户与户之间形成泾渭分明的楚河汉界。然而在划分界线的同时，也割裂了人与人之间的联系交流，且也浪费了不少精力来建立"界线"，在这整体大环境的氛围及涵构脉络中，真正合院精神的建筑简直是难以实现的奢侈要求。
于是，我们翻转了集合式住宅小区当中"界线"的思维，以错落的手法规划出一个近似"米"字的生活场域，取代"井"字的隔离观望，而错落间所产生不同尺度的"缝隙"，将有效让人、动植物、风、味道从身旁的山间渗透流动进来，创造出与环境联结的公共半户外空间。再者，这个小区有个潜规则，每户人家除了拥有自己独立的门前小院外，同时每两户单元间都共享着一个可供栽种的院子，两户人家需商讨后共同经营这个半户外空间；这里的住户有许多的外国籍朋友，不特定的聚会散布在整个小区的角落，成为习以为常的事件风景（Event）。此外，我们让建物间的缝隙也游移至二楼，有如天桥般串连各单元及身旁的地景，成为人们交流互动的绝佳场域。在这里我们想消除公私领域的绝对壁垒，并透过设计的手法植入一个以共享经济（Sharing economy)为基础的小区生活理念。

瓦解壁垒的设计策略

首先将量体碎化，再透过交错的手法创造出模糊性和联结感。最后置入轻质的大平台，创造出可以恣意游走的立体开放生活场域。

渗入透出的缝隙

透过8个"盒子"的错落，创造了许多的"缝隙"，尺度大一点的为小区的小巷弄，成为人们交流来往的场域，尺度小一点的则成为观景停留的个人小角落，在更小一些的缝隙，则常会不经意感受光和风的流窜。透过这些缝隙的延展，也让小区和外部环境有了最频繁的联结，也模糊了内外公私那绝对的分水岭。

过往的历史轴线

我们在小区的主立面设计了一个入口，入口则连接了微型小区最主要的"巷弄"。因为这里曾是过往历史的轴线，所遗留下来的合院并不多，我们只能以隐喻的轴线和红砖来响应，暗示过去的痕迹。

聚落中的个体单元

微型聚落的构成，是我们将每户视作为"个体单元"，群聚错落于彼此间形成"聚落"。每个个体单元都有属于自己的入口，即私人领域，同时与邻房之间会有一处需共同经营的绿地，即共享经济（Sharing economy)的公共领域，模糊了公私领域的界线。为了确保每户仍保有属于自己的入口、视野和景色，所以平面部分会因为每户位在不同位置而有不一样的设计。

代谢的永恒 实现恒久持续的幸福感

另外值得一提的是，本案因为预算及维护的考虑，发展出一套外挂式（Plug-in)的混凝土墙系统，所有混凝土所需要的砾料由现场直接获取，并运至工厂预灌。经过精确计算后的混凝土墙块（Prestress Concrete)，最后再挂至现场的钢结构，大幅降低工程时间与人力成本。同时这种工法也创造出材料具有可替换性的机会，未来任何外墙损伤或因机能需求而调整，都能实现局部更换，赋予生命周期约35~50年的混凝土材料拥有永续代谢的可能，让这座微型"聚落"能真正永续经营。

外挂式（plug-in）混凝土墙系统

为了增进施工效益和节省建造成本，本案研发了预铸外挂式混凝土墙系统，同时也具有可局部更换的优势，以便于维修保养，更为永续建筑迈进一大步。

Constructions for Tiny Urban Tribes

In a densely populated contemporary city, all the city planners and architects reflect, from time to time, how to make an effort in order to gain more living spaces. This difficulty has been temporarily relieved by reinforced concrete high-rise buildings advocated by Domino System in last century of modernism. For that reason, our cities then become a concrete jungle in international style. With a detailed project of space, building coverage and ratio of public utility, a huge and high-density community with multiple-family residential house comes to the world. It became a typical view for most of developed cities. After that, it seems people's life and thought are restricted to a limited space as they established boundaries with neighbors. Nowadays, people are still missing the large garden area of courtyard houses in countryside in old times. On a grain-and-dry-vegetables sunning ground, children played with each other under the sun, and adults sat, talked with each other, played Chinese chess and drank tea under the sun shelter. Then we ask ourselves, can't this scene be realized at present?

A Modern Courtyard Settlement Within a Concrete Forest

The site is located in a rural area of Baoshan, Hsinchu, surrounded by nature. It is conveniently situated in close proximity to the Hsinchu Science Park and the entrance tunnel to the site is most adequately encased with bamboo forests and greeneries. This project involves the renovation of old architecture into a new form of community settlement. We hope to bring together traditional congregated housing community and traditional courtyard to create a rich and diversified living environment that suffices both natural and lifestyle factors.

Wobbling Splits - Living Concept of Sharing Economy

This is a project specializing in tiny tribes. We contemplate the possibility of combining management concepts in modern community and community spirits of traditional courtyards. The main disadvantage of multiple-family residential house community is the indifferent boundaries among households. Once the boundaries are established, they stop the connection and interaction among people, and also waste an amount of time and energy setting them. In an environment with these circumstances and context, it is undoubtedly hard to realize the community spirit of traditional courtyards.

Therefore, we overturn the boundary concept in multiple-family residential house, and we create a living area with a shape of star key on the phone in a scattered manner. It replaces the shape of pound key leading to isolation. And the splits in different scales caused by scattered style can make people, animals, plants and wind in the mountain flow into the area, which creates a public semi-outdoor space connecting with the natural environment. Moreover, the community with this style has an implicit rule; that is, each household possesses not only an independent little garden but also a planting field shared by two households. These two households need to discuss how to deal with this shared semi-outdoor space. In the meantime, there will be many foreign people living in and non-specific gatherings taking place in the community. In that way, such event will become a habitual scene for everyone. More and above that, we move the splits of building to the second floor, which connects all the units and scenes, will be the best place for people to interact and communicate with each other. We are going to disperse any public and private boundaries in this place and sketch a fundamental community living concept of sharing economy.

The Design Strategy to Disperse the Boundaries

First of all, we crumble down the building mass to create an obscurity and an interconnection in a scattered manner in the area. After that, we lay a large light platform on in order to make a three-dimensional open space for people to walk as they please.

The Splits of Flowing Space of Natural Environment and Concrete Constructions
Eight scattered boxes will create many «splits».

Splits in large dimension are those lanes in the community which provide people with a space to come and go. Splits in small dimension are the personal corners for people to stay and view over the scene. There are even smaller splits at where we feel the light and wind from time to time. Through the extension of these splits, we connect, in a frequent way, the small region and the outside world and blur the boundaries between private and public and also internal and external distinctions.

Line of History in the Past
We design an entry at the main facade of community. This entry links the main lanes and streets in the tiny community. Because it was a line of history in the past, there are no more left from the traditional courtyards. We can only utilize implicit lines and red bricks to show the tracks in the past.

Individual Units in the Settlement
For the structural basis of the micro–settlement, each household is viewed as an "individual unit" and units are intertwined in to a community to form a "settlement". Each individual unit has its own entrance and private area. There is also a green area that is managed by two neighboring units, realizing the concept of sharing economy in the public space and clouding the boundary between public and private spaces. To ensure that each household has its own entrance, view, and scenery, the floor plan designs differ depending on the location of each unit.

Eternal Metabolism Realizes a Sustainable Happiness
It's worth mentioning that, due to the consideration for budget and maintenance, we develop a plug–in concrete wall system. All the materials that need for fabricating the concrete are on site. We transport these materials to the factory and turn them into the prestress concrete after an accurate evaluation. After that, we put them into the steel structure in order to reduce the working hours and labor costs. At the same time, this technique creates an opportunity for the materials to be interchangeable. If there are any demands for adjusting the functions or repairing the damage of the wall, it is possible to realize the partial change. The concrete materials with a life cycle from 35 to 50 years are very possible to realize a sustainable urban metabolism. All this can make this tiny tribe finally achieve a real sustainable development.

Plug-in Concrete Wall System
In order to improve construction efficiency and reduce construction costs, this project developed a pre–casted, plug–in concrete wall system that simultaneously bears the additional advantage of interchangeability to facilitate maintenance. This is also a major strive toward architecture sustainability.

NOMINATION FOR THE BEST GREEN ARCHITECTURE DESIGN

最佳绿色建筑设计奖提名奖

武汉市中创环亚建筑景观设计工程有限公司（中国武汉）

Wuhan Zhong Chuang Huan Ya Architecture Landscape Design Co., Ltd (Wuhan, China)

获奖项目/Winning Project

乡园
Country Home

设计说明/ Design Illustration

景观是所有自然过程和人文过程的载体。
"礼失求诸野"，乡村在人类生存空间快速都市化过程中依然是诸多都市情感的缘起和归宿。
设计师用简单的工具例如钢笔来采风和记录稍纵即逝的灵感，用简洁的结构线来还原乡村所拥有的基本景观要素：林地、聚落、河流、田野以及从精神到日常生活的典型时空元素。线性结构勾勒出民居的限定元素，如门窗、墙屋面。天空乃至虚空既是画布的留白亦是变幻的围合实体，虚实相生。
错落有致、斑驳陆离、光影变幻的屋面限定了既是室内又是室外的廊道空间。人们在其间穿行、坐卧、流连或沉思。

Landscape is actually a carrier of all natural and human processes.
"Lost etiquette should be found from the folk". So country is still the origin and home of many urban emotions during rapid urbanization in human living space. The designer is good at drawing and

recording quickly the scene and the inspiration which disappears easily by using simple tools such as pens. Therefore, the simple lines can revert all the basic elements in landscape, including woodlands, villages, rivers, fields, and the ones from consciousness to everyday life. The linear structure depicts the restrictive elements, such as doors, windows, walls and roof.

The sky or void is the blank on canvas, and also the changing substance outside the frame, the false and the true one are always together.

The roof which is full of rich level and varied light defines the indoor and outdoor of the corridor.

Visitors can walk, run, sit, lie, enjoy or contemplate freely in the park.

130
The
Best
Digital Architecture
Design Award
最佳数字建筑设计奖

艾特奖

最佳数字建筑设计奖
THE BEST DIGITAL ARCHITECTURE DESIGN AWARD

INTERNATIONAL
SPACE DESIGN
AWARD

获奖者/**Winner**
Marco Poletto
& Claudia Pasquero
(英国/UK)

获奖项目/**Winning Project**
H.O.R.T.U.S ZKM

获奖项目/Winning Project

H.O.R.T.U.S ZKM

设计说明/ Design Illustration

H.O.R.T.U.S将城市范围的概念理解为扩大的生物圈；由一种新奇的生物数字花园原型，去提供人类消费的可再生能量和营养物；将柱子改造为一种高密度的光生物反应器，利用微藻生态技术，能将人类的新陈代谢和生命的繁殖联结到一起。
能量流动（光辐射），物质（蛋白质），（二氧化碳）和信息（数据）是实时过程和反馈，刺激混合自助组织演化出新形式以及自我规范的新机制的发生。网络园丁受邀直接参与H.O.R.T.U.S项目，以丰富生物数字微生态的物质经验以及表现未来城市网络种植的实践。

H.O.R.T.U.S. Karlsruhe engages the notion of Urbansphere as an augmented biosphere; the synthesis of renewable energy and nutrients for human consumption is reconsidered as an urban practice enabled by a novel bio-digital gardening prototype; the architectural apparatus transforms the archetype of the column into an high-density photo-bioreactor able to connect in space and time human metabolism to the proliferation of life within micro-algal ecologies such as cianobacteria cultures.
Flows of Energy [light radiation], matter [proteins, CO2] and Information [data-feeds] are processed and fed back in real-time, stimulating the emergence of multiple mechanisms of self-regulation and evolving novel forms of hybrid self-organisation. Visitors turned cyber-gardeners are invited to engage directly with H.O.R.T.U.S enriching their material experience of bio-digital micro-ecologies and embodying future urban cyber-gardening practices.

获奖评语

这个项目采用了一种很好的形式，以诗歌般的新设计去塑造一种与自然联系的新形态。人文建筑与自然之间没有距离。

This project is presented in a really good way. It creates a new form to connect with nature through poetic new design. There is no distance between humanity architecture and nature.

NOMINATION FOR THE BEST DIGITAL ARCHITECTURE DESIGN
最佳数字建筑设计奖提名奖

Rafael Beneytez Duran & Ophelia Mantz (美国/US)

获奖项目/Winning Project

Tobogan住宅
Tobogan House

设计说明/ Design Illustration

Tobogan住宅是一座通用的三层楼建筑。该项目两栋房子并列，中间是中空。住宅包含一个停车场、一个入口和两个楼梯。住宅有热调节功能。当漫步穿过长廊、桑拿健身中心、酒吧、餐厅和厨房时，住宅便构建出一组在花园、疯狂和美丽洞穴之间的长镜头。

其中一栋住宅深植于地面，另一栋则脱离地面搭建于空中，是花园高度的两倍。两栋住宅在用材、形状和重量上都完全不同。空间的复杂性则包括两栋住宅所展现出来的各式各样的、模糊的、膜状的空间。

住宅使用混凝土建造，用以承重，并支持侧向荷载，同时利用热惯量控制室内环境对天气变化的反应。

这个项目最特别的地方在于运用了3个相切的圆柱体，它们表面相连。圆柱体的表层使用了一种多层膜。这种膜通过周围环境的刺激进行呼吸：太阳辐射、像液体一样的光、从高速公路和学校传来的噪音。这种铝制微穿孔的板材有助于减少和降低太阳辐射，同时反射回去，并且抵挡噪音。

The Tobogan House is a generic three-storey house on a southern slope. The project juxtaposes two houses and a

void that houses a car park, an entrance and two staircases, working as a thermal regulator and constructing a tracking shot between a garden, a folie and a picturesque grotto while crossing through the lounge, sauna–fitness center, bar, dining room and kitchen. One of the houses is organized to inhabit the soil with its habits and routines. The other was made light-heartedly, emancipated from the ground, in the air, is flying at a double height over the garden. Both are totally different each other in materiality, shape and weight. The most important spatial interest of the complex is the heterogeneous, ambiguous and membranous space that appears between the two houses.

This one is constructed in concrete to preserve the heaviness, address the needs to support the lateral loads of the soil, and make use of thermic inertia to control the environmental reaction of the season weather changes.

This skin is designed as a membrane made by multiples layers which wants to breathe by representing the surrounded environment: sun radiation, light as a fluid, and the impact of the noise over the building which is coming from the highway and a school. So the aluminum micro–perforated panels are useful to decrease and reducing the radiation received and reflected, to break the impact of the noise over the building and to reduce the radiation received and reflected; the Mirror is working to decrease the impact of the sun radiation and to conduct the light as a fluid reflected by the mirrored edges of the windows.

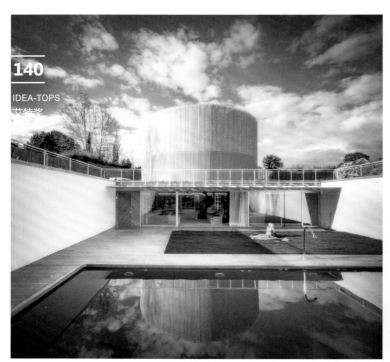

NOMINATION FOR THE BEST DIGITAL ARCHITECTURE DESIGN

最佳数字建筑设计奖提名奖

Volker Giencke (奥地利/Austria)

获奖项目/Winning Project

大琥珀-音乐厅
Great Amber - Concert Hall

设计说明/ Design Illustration

琥珀有一种迷人的透明，特别当它吸入一只小虫时——好像它在保护和照顾一个小生物。通过对比，音乐厅中空夹层的外立面形成一个包层，各种功能被植入这个空间：宏伟的音乐厅、内庭音乐厅、实验舞台、音乐俱乐部、音乐学校和利耶帕亚人民表演场所等。

除了设计独特的建筑和在音乐厅为古典音乐会装备最好的音响设备之一，我们想通过音乐厅给利耶帕亚和它的居民创造一种新的文化认同。音乐厅的建成标志着一种代表其文化认同的新文化产业区的诞生。这是一个独特的、历史性的事件。在建筑和室内设计都被充分认可下，音乐厅项目与城市联结到一起。它是利耶帕亚的一个新的现代标志性建筑。

Amber is intriguingly transparent, especially when it encapsulates an insect – as if it were protecting and caring for a living thing. By comparison, the double-skin façade

of the concert hall forms an envelope that provides a microclimate in which the different functions are included as spatial implants: the grand concert hall, the chamber music hall, the

experimental stage, music club, and music school, etc. as well as the "Civita Nova" as a performance venue and stage for the people of Liepaja.

Besides designing unique architecture and equipping the hall with one of the very best acoustics for classical concerts, it had always been our ambition to give Liepaja and its residents a fresh cultural identity with this new concert hall. The completion of the concert hall's construction marked the birth of a new cultural quarter, of its cultural identity. It was a unique, historic event. Convincing in both architecture and content, this symbolic effect emphasizes Great Amber's connection to the city. It is a new landmark of modern Liepaja.

NOMINATION FOR THE BEST DIGITAL ARCHITECTURE DESIGN

最佳数字建筑设计奖提名奖

李麟学（中国苏州）
Li Linxue (Suzhou, China)

获奖项目/Winning Project

青岛嶺海温泉大酒店
Qingdao Linghai Spa Hotel

设计说明/ Design Illustration

项目强化以体验为主导的五星级酒店设计趋向。
"环境响应建构"是青岛嶺海温泉大酒店设计关
注的焦点。环境参数化设计试图提供一个当代建
筑实践中的路径与范式。设计由环境参数的理性
分析与捕获入手，以热力学能量形式的视角，在
风、光、热、景观诸要素的参数作用下，提出了
沿海面展开最大景观面，并塑造与风的流动、光
的引入响应的如波浪般流畅起伏的总体形态。建
筑曲面空间表皮与内部流动空间，成为设计建造
的一大挑战。曲面玻璃与斜梁构成了酒店激动人
心的峡谷般的中庭空间。

This project strengthened the trend of
experience—oriented five-star hotel
design. The design mainly focuses on
"environmental responsive construction".
The environmental parametric design
tries to provide an integrated paradigm
for the contemporary architectural practice.
Based on rational analysis of
environmental parameters (wind, light,
heat and landscape), the designer tried
to shape the building according to
thermodynamic energy
theory. The building
captures a panoramic
view
of the seaside, and
introduces wind and
light to create a wavy
effect of the
architectural form. It
was a great challenge to
create this curved
spatial facade and to
construct the flowing
interior spaces. Curved
glasses and inclined
beams together
shape the canyon-like
atrium space of the
hotel.

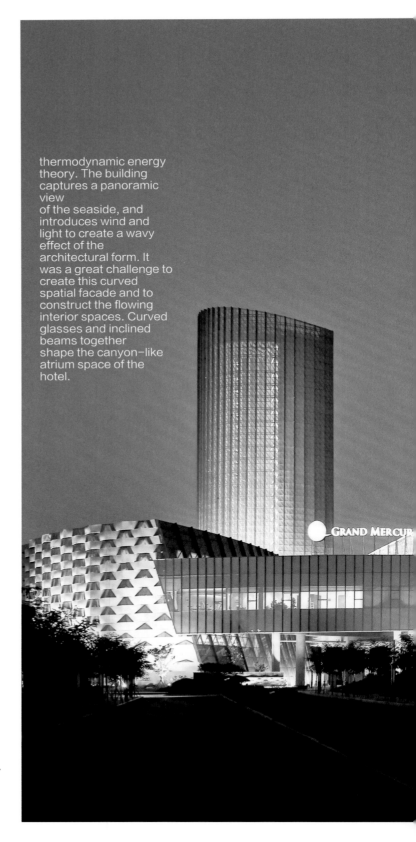

NOMINATION FOR THE BEST DIGITAL ARCHITECTURE DESIGN

最佳数字建筑设计奖提名奖

Marco Poletto & Claudia Pasquero (英国/UK)

获奖项目/Winning Project

城市愚钝的藻类
Urban Algae Folly

设计说明/ Design Illustration

城市愚钝的藻类是一个交互式、整合微海藻文化、未来生物数字建筑的项目实例。微藻，在此实例中为小球藻属，是一种特殊的光合机器。它包含人体需要的基本营养物，如矿物质和植物蛋白。相比大型树木而言，微藻能数十倍更有效地氧化城市大气层的空气并吸收二氧化碳。

城市愚钝的藻类源于众所周知的ETFE建筑表层系统的演变。在此实例中，它可以为刺激小球藻生长以及确保访客的互动提供理想的栖息地。在任何时刻，城市愚钝的藻类那有效的半透明性、颜色、反射率、声音和生产力，是气候、微藻、人类和数字控制系统共生关系的结果。

The Urban Algae Folly, is an interactive pavilion integrating living micro–algal cultures, a built example of architecture's bio–digital future. Microalgae, in this instance Chlorella, are exceptional photosynthetic machines; they contain nutrients that are fundamental to the human body, such as minerals and vegetable proteins; microalgae also oxygenate the

air and can absorb CO_2 from the urban atmosphere ten times more effectively than large trees.

The innovative architecture of the Urban Algae Folly originates from the evolution of the well known ETFE architectural skin system; in this instance it has the ability to provide the ideal habitat both to stimulate Chlorella's growth and to guarantee visitors interaction. In any given moment the effective translucency, the colour, the reflectivity, the sound and productivity of the Urban Algae Folly are the result of the symbiotic relationship of climate, microalgae, humans and digital control systems.

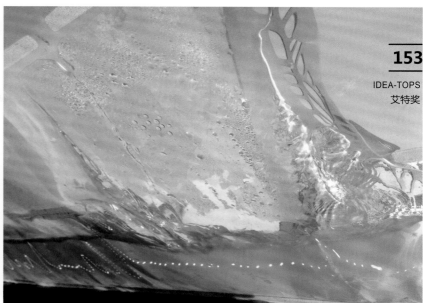

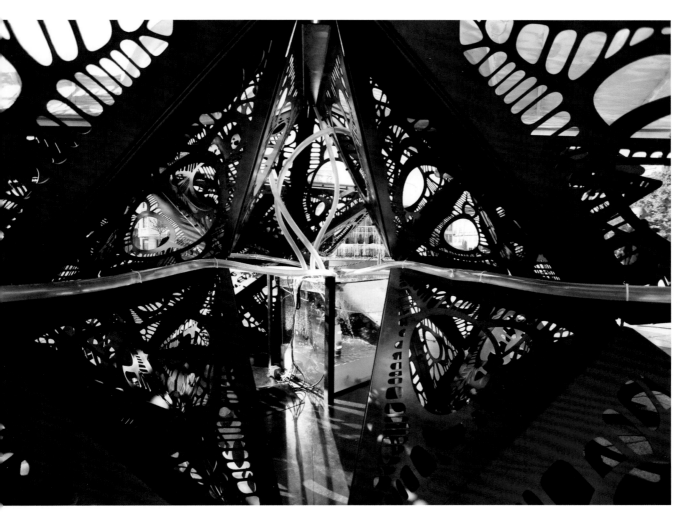

154

The
Best
Commercial
Space Design
Award

最佳商业空间设计奖

艾特奖

最佳商业空间设计奖
THE BEST COMMERCIAL SPACE
DESIGN AWARD

INTERNATIONAL
SPACE DESIGN
AWARD

获奖者/ Winner

郑州筑详建筑装饰设计有限公司（中国郑州）

BDD.HK (Zhengzhou, China)

获奖项目/Winning Project

康桥悦城生活美学生机馆

Sales Center of Joyous Family

获奖项目/Winning Project
康桥悦城生活美学生机馆
Sales Center of Joyous Family

设计说明/ Design Illustration

自然、艺术&家
自然是生命的载体，家是生活的容器，而艺术是为鲜活日子注入的氧气。

低头族群在全球范围内扩张蔓延，在我们还未觉知之时，已经成为其中的一员了，它的症状特征是所有患者的零碎时间全被以强制低头的姿势消耗殆尽，人与人之间原有的多样的情感传递方式也被迅速改变为单一与单向的。从我们互动的方式到我们居住的方式都在迅速发生着变化，从小时候的院子迁徙到高层建筑的格子房，从融入自然到与自然割裂，快节奏的生活迫使身心处于亚健康状态，麻木我们的感知力，特别是感知自然与幸福的能力。人们的身心该如何调节才能匹配当下的生活状态，什么样的环境设计能改善低头族群的症状，缓解人与人之间的互动危机，这正是从事空间设计的我们应该思考与研究的核心问题。

地产商是以家为核心的生活缔造者，是生活方式及生活美学的传播者。营销中心是能唤醒客户对家的所有想象及幸福感的空间场域。模糊营销中心的单一功能，融入美术馆、生活馆、图书馆、自然生态馆等功能的互动与体验，可以帮助地产商向客户展现他们不仅立足商业，还立足替客户研发未来健康生活模式的意愿与成果展现。本案运用植物墙及自然生长的竹林隔断，来淡化室内外空间之间的界限，并且这种方式延续到各功能区域之间的划分。通过这种方式来打破建筑六面体对空间的约束，从而营造轻松自在、生机盎然的体验氛围。再结合陈设品情景化的转变，来营造情感上的互动体验氛围。这些体验氛围会和我们儿时的记忆相连接，试图唤醒对大自然及过往生活场景的回忆。譬如，大面积的墙壁及地面留白，试图留给思考当下生活状态的艺术家，聚焦他们敏锐的作品来引发观者的共鸣与思考。又如，堆积的格子替代书柜的隔断功能，里面摆放了家喻户晓的老书籍，儿时陪伴的物件，以及碧绿青翠的植物。体验式厨房、儿童课堂及多媒体体感互动区等功能区域的添设，让观者在回忆过往生活的同时，平添了对未来生活场景的畅想。各种情景的设定，通过五感的体验、觉知，连接了观者的情感，锁定了心灵的所想与所向，在互动中悄无声息地传达着细腻人本的生活美学。

Nature, Art & Home
Nature is the carrier of life, home is the container of it, and art is the oxygen which vivifies it.

Phubbing is spreading across the world, and we already have this habit before we became aware of it. Its symptom is that the patient would consume all the spare time lowering his/her head and playing mobile phone or other electronic devices. People used to have different ways to express their emotions, but now they tend to use a single and unidirectional way. The ways of interacting and living happen to be changing so fast. From the yard in our childhood to the apartment on high buildings, and from being a part of nature to a separation from nature, the high pace of life has pushed our mental and physical state into a sub-health state, and weakened our perception, especially our ability to access nature and happiness. How our heart and body should adjust to keep in pace with the way we live, and what environmental design can ameliorate our symptoms of smartphone addiction and soothe the interaction crisis between individuals, for people who work on spatial design, these are the core questions worthy of research and contemplation.

A land agent should be a life maker, who focuses on home and transmits the way and aesthetic of life. A marketing center should be a space, where could evoke a sense of happiness and all the imaginations about home. By blurring the singular purpose of marketing, and providing functions like fine arts museum, life museum, library and ecology museum, we could help land agents

demonstrate not only their position in business, but also their will to research and develop a healthy style of life in future from the perspective of their customers.

This project uses plant wall and groves of bamboo to separate and blur the boundary between inner and outer spaces, and extends this application to all separations of areas with various functions. In this way, we could break the restrictions a constructional hexahedron has imposed on space and foster an animated and relaxed atmosphere; moreover, the furnishings are varied with different scenes, enabling people to experience an interactive atmosphere. The atmosphere would connect our childhood memory, and try to remind us of the scenes of nature and early life. For example, large white spaces on wall and ground are prepared for artists who are focusing their mind on current life state, and their remarkable works could trigger viewers' resonance and contemplation. Moreover, piled squared cells replace bookshelves to separate spaces, and they are filled with well known old books, things that accompanied one's childhood as well as some flourishing green plants. As for functional areas like experiential kitchen, children classroom, and multi-media interactive zones, they could not only draw viewers to their memory of past time, but open their imagination about future life. The design and arrangements of various sceneries, through sensations and experience of five senses, would connect with viewers'emotion and direct where and how their minds shall be. We convey the subtlety of the aesthetic of life and the value of humanity in a silent way.

获奖评语

这是一种商业空间的
新模式，同时在设计
上很讲究文化，在商
业空间中重视文化和
自然的表达。

This is a new
model of
commercial space,
while emphasizing
cultural element in
the design. It
emphasizes the
importance of
cultural and natural
expression in
commercial space.

NOMINATION FOR THE BEST COMMERCIAL SPACE DESIGN

最佳商业空间设计奖提名奖

estudio ALA

estudio ALA (墨西哥/Mexico)

获奖项目/Winning Project

Nail It 美甲艺术工作室
Nail It II

设计说明/ Design Illustration

Nail It是一个位于墨西哥哈利斯科州的美甲艺术工作室。工作室内彩色天花板的设计灵感来自于五光十色的美甲产品的选色板，内部空间被设计成一个既可以被快速复制又不失艺术氛围的商业空间。

色彩鲜艳的灯光装置照亮了底部的工作空间，重叠交互的光影效果使人们进入的时候感觉天花板在不断变幻。接待顾客的空间被垂直分为两部分：地面的活动空间和悬挂于天花板的7000根彩色柱条组成的空间。这些悬挂的装置每一根都彼此独立而又形成一个视觉整体。从粉色到蓝色，由暖色调逐渐变成冷色调，各个颜色相互衔接，营造出连续性。

底部空间则全部采用了白色，让人感觉井然有序、一尘不染。这一区域就像是一块白色画布，人们在这里进行各种各样的美甲艺术创作。特别设计的家具保证了空间的流畅。工作室的最里面有一个玻璃围成的区域是小孩子的活动空间，母亲可以把孩子放在这里玩耍然后专心做自己的事。彩色的设计概念被应用到空间的各处，儿童区的粉色系塑胶地板采用了和天花板相同的颜色，上下的色彩呼应使整个设计显得独特而浑然一体。

Nail it is a nail-art studio located in Jalisco, Mexico. Making reference to the colorful palette of nail polishing products, the project was designed with a strong palette to create an identity that could be replicated in any commercial space without losing the feeling of the studio.

The light and color installation provide illumination into the customer area, provoking diverse textures and effects when the hanging sticks virtually overlap and give the feeling that the whole ceiling changes as you walk through the studio. The customer area works as an open floor plan and is segmented vertically in 2: the client area and the ceiling installation with 7000 colored sticks. These hanging elements are attached individually and work as a whole. The colors chosen for the ceiling intervention have a relation between each other in the color diagram, these help the whole concept to transmit the whole idea of continuity. These specific 4 colors start with pink were the warm colors start to turn into the cold color area.

On the lower client area everything was maintained white to transmit the tidy filling of the studio where the hygiene is flawless, this area serves as a white canvas for the wide variety of designs in nail art possible. Specially designed furniture was taught to maintain the feeling of continuity into the space.

On the back of the studio, enclosed by clear glass, one can find a children's area where kids can enjoy themselves while their mothers are being attended by the staff. The area is defined by a pink plastic floor in the same color as the ceiling, giving continuity to the whole concept. It is in this area were the ceiling intervention reveals its specific yet universal condition.

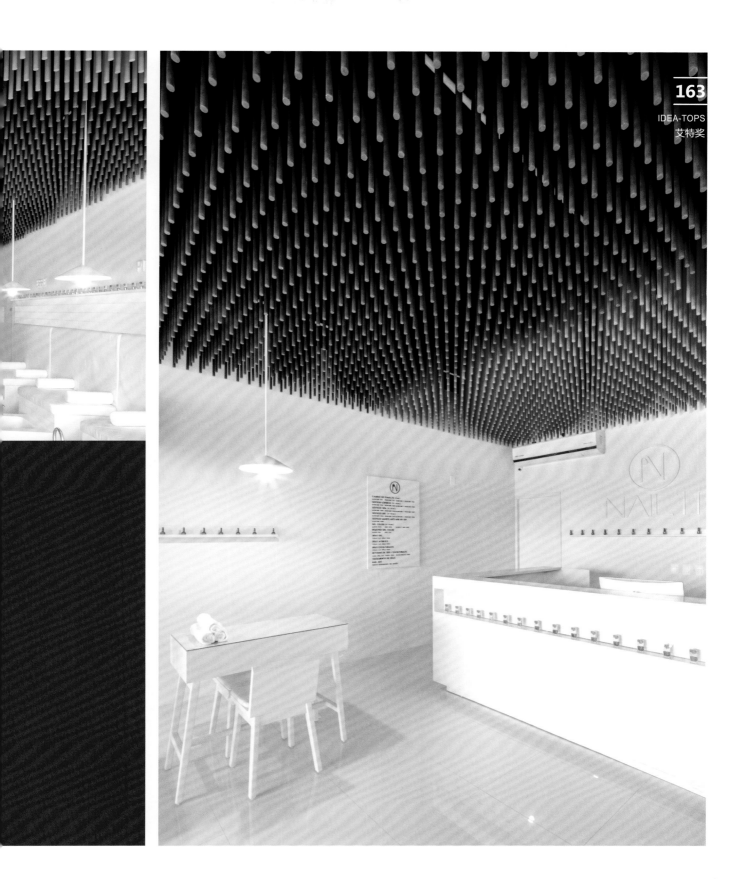

NOMINATION FOR THE BEST COMMERCIAL SPACE DESIGN

最佳商业空间设计奖提名奖

万浮尘（中国苏州）
Fc Wan (Suzhou, China)

获奖项目/Winning Project

青岛东方时尚发布体验中心
Qingdao Oriental Modern Experience Center

设计说明/ Design Illustration

本案位于山东省青岛市黄岛区。是集T台秀场、大师作品展、接待中心、创客服务空间、会议、办公为一体的多功能体验中心。
整个空间设计是延续建筑的流线形设计语言来展开，为了加强室内空间与整个建筑相互呼应关系。室内空间围绕中央共享大厅展开设计，中央区域是一个面朝大海的白色旋转楼梯，白色旋转楼梯俯瞰犹如海边一颗海螺，又似一个妙龄少女穿着时尚礼服在海边起舞。同时旋转楼梯的向上延伸，塑造出了天使般的翅膀，向往着天空飞翔。
中央大厅的顶面，原建筑为玻璃屋面，为了减少能源损耗，采用大面积的双层柔性发光膜来做顶面设计，白天可以过滤掉部分太强的自然光，室内也不至于很暗，晚上利用人造光营造出柔和的顶面发光体，让人有一种要冲到顶面的欲望。蔚蓝的大海、洁白的旋转楼梯两相完美结合，缔造出时尚简洁之美。

Located in Huangdao District, Qingdao, Shandong, this project is a multi-functional experience center which gathers T stage, works exhibition, reception center, service space for makers, meeting and working space.

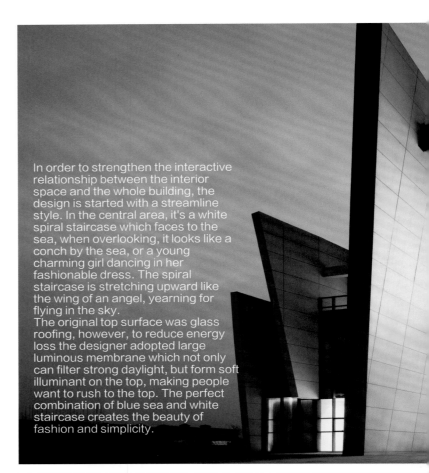

In order to strengthen the interactive relationship between the interior space and the whole building, the design is started with a streamline style. In the central area, it's a white spiral staircase which faces to the sea, when overlooking, it looks like a conch by the sea, or a young charming girl dancing in her fashionable dress. The spiral staircase is stretching upward like the wing of an angel, yearning for flying in the sky.
The original top surface was glass roofing, however, to reduce energy loss the designer adopted large luminous membrane which not only can filter strong daylight, but form soft illuminant on the top, making people want to rush to the top. The perfect combination of blue sea and white staircase creates the beauty of fashion and simplicity.

NOMINATION FOR THE BEST COMMERCIAL SPACE DESIGN

最佳商业空间设计奖提名奖

深圳市朗图设计有限公司 (中国深圳)
Shenzhen RITO Design Co. Ltd.
(Shenzhen, China)

获奖项目/Winning Project

前海壹会
Qianhai ONE S

设计说明/ Design Illustration

壹会品牌哲学源自"一期一会，难得一面，世当珍惜"，因而竭尽全部心力准备好每一次亮相，为了人生中可能仅有的一次相会。壹会，希望创造一个别具一格的空间，诠释人与人，人与生活，人与艺文的完美邂逅。
以城市多元文化生活为起点，以当代性的美学观，将艺术、文学、佳肴、美酒期会于此，兼容并处，并通过各种艺文活动活化场所精神，让每一位参与其中的都能成为壹会内涵的共同缔造者，分享美学，启迪思想。
从市场和经营出发，我们将壹会定位为集艺术、美酒、美学、生活为一体的中国核心城市商业主力店旗舰。以"当代·繁盛·愉悦·极致·定制"为核心价值，组合悦食、艺文、博物、欢饮成为全新业态，将壹会打造为当代生活美学的理想范本。

The philosophy of ONE S origins from "If we could only see each other once during our lifetime, we shall cherish it very much", therefore we put all of our efforts to prepare every design for possibly the only meeting in life. ONE S hopes to create a special space to interpret the perfect meeting between people, people and life, people and art as well as literature.

Starting from diverse cultural life in city, based on contemporary aesthetic view, the project gathers art, literature, delicious food and wine together. Everything is compatible with each other. The site spirit is activated by all kinds of art and literature activities, and everyone participates in could be a co-founder of the connotation of ONE S, sharing aesthetics and enlightening thoughts.

From market and operating point of view, we orient ONE S as the main commercial flagship store in core cities of China which gathers art, literature, good wine, aesthetics and life. The core value is "contemporary, flourishing, joyful, extreme, custom", combining delicious food, art and literature, abundant materials and joy as a brand new type of operation. All of these make ONE S an ideal project of contemporary life aesthetics.

[""]

NOMINATION FOR THE BEST COMMERCIAL SPACE DESIGN

最佳商业空间设计奖提名奖

李超、朱毅（中国福州）
Li Chao, Zhu Yi (Fuzhou, China)

获奖项目/Winning Project

Forus

设计说明/ Design Illustration

业主"Forus"为高端定制的婚纱机构，设计师根据该婚纱机构的针线及蕾丝等元素，结合了建筑loft的工业风格，思考如何将蕴含于空间内的空间本质挖掘而出，设计师最后将糅合后的婚纱浪漫感性元素及loft工业风的粗犷硬朗元素散碎在空间中，两者结伴同行相映成趣。

整个空间中门头的立体钢架及内部钢架的结构，通过不同类型的玻璃——钢化玻璃及镜子的穿插运用，配以蕾丝花纹的墙纸、混搭抢眼的花砖，将空间封装在其中。立体干净的结构是该空间诉说的主题，个性斑斓的沙发组合、刚毅别致的灯具、精致高雅的婚纱仿佛是该主题中的各种花絮，让每一个空间都成为一个独特的视觉机制，折射着、透视着、跳跃着每个参观者对婚纱别样的梦。

"Forus" is a high-end custom wedding dress agency, the designers were thinking about how to dig out the space essence which inherent in the space by combining with the industrial style of loft building and according to the sewing and lace elements of the agency. Finally, they scattered the blended romantic and emotional elements, and the rough and tough elements of loft's industrial style in space, forming an interesting contrast.

The three-dimensional steel structure of the door head and the inner steel structure, through interspersed application of tempered glasses and mirrors which are matched with lace wallpaper and eye-catching tiles, encapsulate the space in them. The clean three-dimensional structure is the theme of the space, the personalized and gorgeous sofas, steely and chic lamps, exquisite and elegant wedding dresses seem to be the highlights which create unique visual features for every space, where every visitor's specific dream of wedding dress is refracting and dancing.

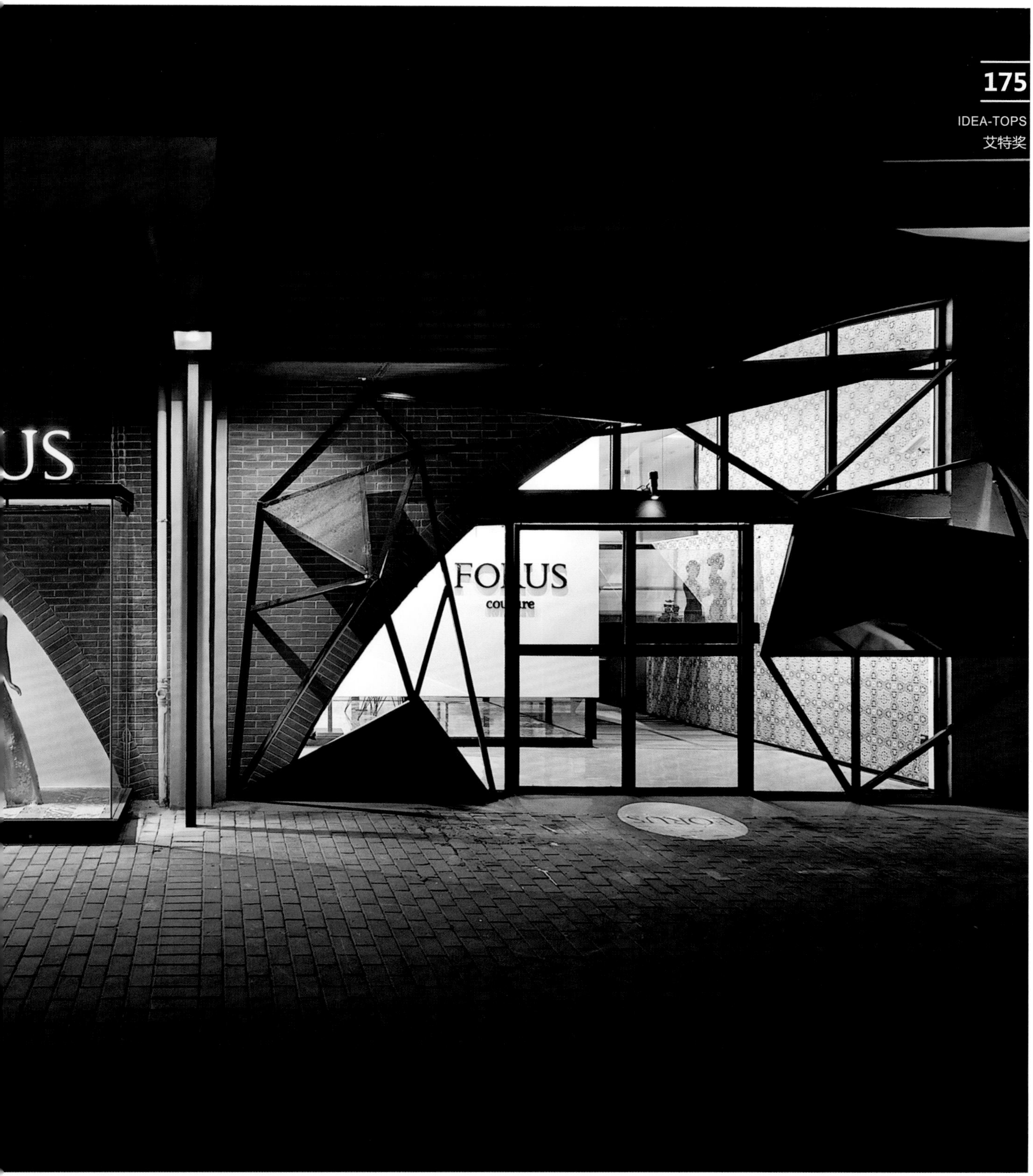
FOLUS
couture

178

The Best
Cultural
Space
Design Award

最佳文化空间设计奖

艾特奖

最佳文化空间设计奖
THE BEST CULTURAL SPACE DESIGN AWARD

IDEA TOPS

INTERNATIONAL
SPACE DESIGN
AWARD

获奖者/ **Winner**

BAUKUH

(意大利/Italy)

获奖项目/**Winning Project**

米兰回忆之屋/Milan House of Memory

获奖项目/Winning Project

米兰回忆之屋
Milan House of Memory

设计说明/ Design Illustration

"回忆之屋"是米兰的一个档案馆、展览空间及会议空间。它是意大利5个文化协会为了纪念获得自由和民主而设立的总部，这5个文化协会分别是：被驱逐者国家协会（A.N.E.D.）、意大利游击队国家协会（A.N.P.I.）、恐怖主义受害者国家协会（A.I.VI.TER.）、1969年12月12日丰塔纳广场协会及意大利抵抗运动国家研究所（I.N.S.M.L.I.）。该项目在2011年以一等奖获得建筑方案竞标，并于2015年完工。

"回忆之屋"的结构非常简单：它就是一个长20米、宽35米、高17.5米的"盒子"。整个建筑被划分为3个区域，区域之间通过底层的开放式结构相互连接。较低的两层是档案室（南边）、休息室和技术设备区（北边）、垂直通道。底层的开放空间被两个八边形的柱子划分为三块，这块区域的三分之一直通顶层，中间放置了一个螺旋梯，其他区域则是展览空间及办公空间。设计师在三层办公和展览空间以及五层档案室的内部嵌入了一个巨大的黄色螺旋梯，使整个建筑显得更加宽阔。档案室的紧密层级与楼梯的巨大体积形成对比，给办公和展示空间带来空旷感，让访客感觉置身于大气、开阔的氛围之中。

黄色的螺旋梯不仅仅是整栋建筑划分区域的主要元素，同时它还在访客和收藏品之间建立了一种联系。由于档案室的物品非常珍贵，访客是不允许直接接触文献的，而市民与藏品通过旋转上升的楼梯产生联系。访客在反复接近或远离藏品的过程中，不仅可以浏览到排列复杂的文献，还可以看到建筑外部公园的景致。

光线以各种各样的形式进入建筑内部。办公区域的大型窗户集中在与建筑外立面相对应的位置。整个建筑运用了开放式的布局，同时内部使用玻璃进行分隔，使室内光线充足。入口大厅、楼梯以及底层空间则从少数几个大型开口处获得光线，微弱的光线进入到昏暗的空间中，营造出静谧、庄严的氛围。

The "House of Memory" is an archive as well as an exhibition and conference space in Milan. It is the headquarter of five cultural associations whose aim is to preserve the memory of the conquest of freedom and democracy in Italy: the National Association of Former Deportees (A.N.E.D.), the National Association of Italian Partisans (A.N.P.I.), the National Association of Terrorism's Victims (A.I.VI.TER.), the Association Piazza Fontana 12 December 1969 and the National Institute of Studies on the Italian Resistance Movement (I.N.S.M.L.I.). The building was completed in 2015, on the base of the project awarded first prize in an invited architecture competition organized in 2011.

The House of Memory is a very simple building: it is a box with a rectangular base of 20m by 35m and 17.5m high. The building is divided into three parts that are connected to one another by an entirely open ground floor. Two thin layers along the building's shorter ends house the archive (South), the restrooms and technical installations (North), and the vertical circulation. The open space at the ground floor is subdivided in three parts by two octagonal columns. One third of this area reaches the building's full height and includes a spiral staircase. The rest is occupied by exhibition spaces and offices disposed on three levels. This internal organization—with the enormous, yellow staircase inserted between the three levels of offices and exhibition spaces and the five levels of archive—introduces a greater scale into the building. The contrast between the tight levels of the archive and the colossal dimension the staircase allows the office and exhibition spaces to acquire spaciousness; the visitor perceives a vaster, more generous atmosphere.

The yellow staircase is not only the building's main distributive element, but it is also the device that establishes a relation among the visitors and the collection. Given that the preciousness of the archive does not allow the visitors to directly access to the documents, the relationship between the citizens and the collection is established through the rotating movement created by the staircase. The visitor repeatedly comes closer to and then moves away from the collection, thereby experiencing a complex sequence of views of the documents and, behind them, of the park outside.

Light enters the building in many fashions. The offices have large windows concentrated in the corresponding parts of the façade. The building's open-space configuration and internal subdivision with glass emphasize the abundance of light. The entrance hall, the large staircase and the ground floor receive light from a handful of very large openings; grazing light invades the semi-darkness and generates a calm and solemn space.

获奖评语

这个项目很简洁，使用了传统材料，这同时展示了美学和诗意。营造了一种建筑内的空旷感。而这种空旷其实是很充实的展示。

This is a really simple project that uses some traditional materials, which is a really aesthetic and poetic solution at the same time. It creates a sense of void inside the building. So this sense of the vacancy seems to be really present inside the building.

NOMINATION FOR THE BEST CULTURAL SPACE DESIGN

最佳文化空间设计奖提名奖

SoNo Architects

(斯洛文尼亚/Slovenia)

获奖项目/Winning Project

意大利图书中心
Book Center, Italy

设计说明/ Design Illustration

由于奥贝丹广场和法比亚尼设计的国家大厅距离很近，这促使设计公司决定为居住在意大利里雅斯特市的斯洛文尼亚少数民族建立一些象征性的联结，这对他们来说是很重要的。设计公司从法比亚尼设计巧妙的外部装饰中获得灵感，将装饰性的几何元素进行分解，转变成现代感十足的结构，成为书架的主要造型。

图书中心所在的历史性建筑是一座可以追溯至20个世纪上半叶的宫殿。建筑的理念是将书架设计成一道景观，改善现有的室内装饰，为空间注入现代感。书架是室内的主要元素，并且它们对访客和城市来说都具有重要作用。建筑的外立面极富光泽，从柱廊处可以看到图书中心内部的场景。如今，这个斯洛文尼亚的图书中心已经成为城市中非常活跃的一个区域，因此也成为各种活动、工作坊、会议的举办场所，并且对城市的所有居民都是开放的。

温馨的室内氛围吸引着访客一次又一次来访，它不仅仅是一个书店，同时还是一个为形形色色的访客举办不同社会文化活动的场所。年轻人可以在这里享受阅读网页新闻和电子书的乐趣，小孩们可以玩游戏、阅读儿

童读物或在工作坊学习。而年长的访客可以在舒适的座位上休息、阅读，并感受周边的温暖气氛。

一进入室内，访客们就徜徉于书林之中。书架的独特造型搭配交错摆放的作品让其成为空间的亮点。室内主要运用了木材，营造出温润质感，并与明亮、光滑的地板形成鲜明对比，使得木材的颜色和结构更加醒目。

The proximity of the Oberdan Piazza and Fabiani's National Hall has encouraged the design company to look more closely to the symbolic ties that are important for the Slovene minority living in Trieste. A hint that would only symbolically show itself derives from Fabiani's masterfully designed facade's decoration. Decorative geometric element has been broken down and transformed into a modern structure, which represents itself in the basic pattern of the bookshelves.

The historical building where the new center is located is a palace dating from the first half of the last century. The architectural concept of the interior upgrades the existing facility with a 'scenery' composed out of bookshelves which give the space a modern look. Bookcases are the central element of the interior, announcing its function to the visitors and the city. Large glazed surfaces oriented towards the free-covered colonnade open the views into the interior of the book center, which results into becoming a lively part of the city. City space therefore becomes an arena for different events, workshops and conferences – Slovenian culture is open to the city and its habitants. Interior is warm, inviting visitors to enter and visit it again and again. The book center is not only a bookshop, but a place for organizing social and cultural events for all type of visitors – It thinks of its young generation of visitors who like browsing the web for news and culture and who enjoy e-books. For children, there is a reading corner that provides games, children publications and different workshops. Elderly visitors can rest and browse through the books in the comfortable sitting area while not missing the surrounding atmosphere.

The visitor while entering is surrounded by books. The pattern of the bookshelves and its composition accompanied by interactive displays become the central point of the space. Material that dominates is wood, which exudes warmth especially in contrast with the bright smooth floor. This way the wood's color and structure become even more pronounced.

NOMINATION FOR THE BEST CULTURAL SPACE DESIGN

最佳文化空间设计奖提名奖

Love Architecture inc.

Love Architecture Inc.
(日本/Japan)

获奖项目/Winning Project

龙泉寺纳骨堂八圣殿
Hasshoden - Charnel House in Ryusenji Temple

设计说明/ Design Illustration

将寺庙作为与"生"互动的场所向社区开放
近年来，随着城市化的不断发展，日本人对于宗教的感觉，包括对祖先的尊敬发生了改变，本地社区不断衰落，按照Jidan系统到寺庙注册证明自身信仰的人也在不断减少，削弱了寺庙的财政基础。纳骨堂回应了这一时代趋势，不再遵循Jidan系统，创造了与代代相传的墓地不同的新的寺庙形式。
寺庙位于川口市，是一个四周由高高的围栏围合的封闭区域，其花园被改造成了葬礼大厅和停车场，主殿环境受到了压抑。这里的纳骨堂有超过40年历史，由于鲜少使用，所以被安静地置于一旁。佛教本不是单纯保佑家庭兴旺的宗教，祭奠行为不会轻易随着墓地的形式变化而改变。因此，建筑师决定纳骨堂的改造不仅要与"死"相关，还要与寺庙成为一个整体，让"生"的人也参与进来。
纳骨堂曾经的曲线形表面被改造成了直线，花岗岩让人们把注意力放到了建筑本身。寺庙前的围栏被全部移除，取而代之地种以树木，在视觉上将葬礼和纪念区分隔开来，让传统宗教自然的指引力重新被释放出来。在"回忆之路"的边上，设计了盥洗池、休息室和分岔路，主入口区域用于举行纪念活动，地藏和佛像被聚集在一起沿着入口摆放。

用一次性的自然现象设计祭奠时间
尽管按照预定和谐说，墓地的形式被改为了纪

念碑等形式，但是回忆死者的时间和空间应该是不俗并且超越人类认知的。入口两侧种有花草树木，阳光透过树叶洒下点点光斑，被风拂过的枝叶沙沙作响，花开花败，告诉人们四季的更迭。休息室和盥洗池富有表现力的墙面以及水池中的波纹都成为了自然的滤镜，将一次性的自然现象呈现在人们面前。由此，通往纳骨堂的小路成为了"生"的代表，与建筑内的"死"形成对比，架过水面的楼梯成为了生与死的分界。纳骨堂呈八角形，象征冥府，也暗示着宇宙。入口处昏暗的灯光让人意识到环境的变化，中央为一座素土夯实的旋转楼梯，光从顶部照下，仿佛人们正处于地下，在旋转的过程中不断自省。当你到达顶部，发光的门板上就像有着不同的星球。光线透过竹子的缝隙照射进来，如星光般闪耀，你就是在这里缅怀那些逝去的人。

生死的守望者
改造之后，寺庙方反映说有更多人带着孩子来参观，这正是建筑师所希望的。夏天时孩子们可以在水池里玩耍，在寺庙中的嬉戏成为了他们童年的一部分，让他们长大后也会来到这里，珍惜这里，直到他们结婚生子以至生命结束，寺庙将一直安宁地守望着他们的人生。设计让寺庙回归了本来该有的状态，它应该是永恒不变的。

Opens the temple to community as a place to interact with "life"

In recent years, the decay of local communities due to urbanization has changed the Japanese sense of religion including ancestor worship, weakening the "Jidan" (system where commoners had to register with a temple to prove their Buddhist faith) which is the financial basis of many temples in Japan. Charnel house, which is a new style of grave that appeared as though it responded to the trend of the contemporary times, is different from the graveyard passed down from generation to generation and does not follow the traditional Jidan system.

Amidst such time period, this temple, which stands in the city of Kawaguchi, had the garden changed into a funeral hall and parking lot, oppressing appearance of the main temple building. The tall fences surrounding the premises made the buildings a closed area. The over 40–year–old charnel house, which was rarely in use,

stood in a quiet way. Buddhism is originally not a religion that just supports the succession of families. The custom of visiting graves does not change easily even if the form of graves changes. This is why we decided not only to renovate the charnel house related to "death", but also to reproduce the temple as a whole that involves people with "life".

The originally curvy front approach was made straight, and existing granites were used to focus the perspective on the main temple. The place where it was originally the front approach was planted with trees to visually divide the area into funeral and worship areas. The fences along the front approach were then all removed, releasing the drawing power that a traditional religion naturally has. The locations of washstand, restroom, and the branching approaches that extend from the "Shinobi no komichi" (pathway of recollection) and the front approach were decided in relation to worship activity and the whole lot. Jizo and Buddha stone statues that had been gathered in one place were appropriately distributed along the front approach.

Designs the hours of worship through one-time natural phenomenon

While the forms of many graves have transformed into monuments and mechanism, relying on human-made pre-established harmony, the time and space for recollection of the dead should be something transcendent and beyond human knowledge. The trees along both sides of the front approach and the flowers at one's feet show one the sunlight through trees, and wind through rustling of leaves, and tell the change of seasons through different fruiting and blooming of flowers. The expressive walls of restroom and washstand and the ripples on water basin reflect the one-time natural phenomenon and function as a filter of natural tremor. This way, the path leading to the charnel house was set as a representative of "life" in opposition with the charnel house and graves representing "death". The staircase bridging over the water basin functions as the boundary, which separates life and death. The octagonal charnel house represents the Hades; essentially the form of the universe. The dim lit entrance makes one aware of the change in place through the luminance difference. The spiral staircase in the center made of rammed earth allows the top light from the sky to enter, reminding you of underground and promoting introspection within the rotational motion. When you reach the charnel chamber, the luminous doors look like planets. The light streaming in through the gap in the bamboo ceiling blink like the stars in the space. This is where you face the deceased. The series of architectural facilities appeal directly to your perception along the time sequence of worship.

A fixed point to watch overcycle of life

After the renovation, they say there are more visitors accompanied by children. We wish that children would play in the water basin during summer. They would play in the temple in their childhood. They would visit the temple and cherish it. Then they would marry and have children, and eventually pass away. We hope that this temple would be a place to watch over such cycle of life peacefully. The plan was to reconstruct the temple into its original state. Temples should remain the same.

NOMINATION FOR THE BEST CULTURAL SPACE DESIGN
最佳文化空间设计奖提名奖

刘锐韶（中国佛山）
Liu Ruishao (Foshan, China)

获奖项目/Winning Project

广东锐美集思创新中心
Guangdong RUIMEIJISI Innovation Center

设计说明/ Design Illustration

位于中国广东工业设计城的锐美集思创新中心是一座由旧建筑改造而成的设计作品，旧建筑前身是一座给800多位设计师提供午晚用餐的设计师食堂。从2012年到现在，设计城已发展到近2000名设计师，新的食堂启用后，旧食堂功能变为设计创新中心。本案采用了一种新东方的设计方式结合国际简约时尚的审美趋向，将建筑物的室内外进行重新设计打造，这种新东方风格是传统与现代的有机结合，是中国传统风格文化意义在当前时代背景下的演绎，是在中国文化充分理解基础上的当代设计，是对传统文化的合理继承与发展！

室外的原有树木得到设计师通过巧妙的设计而保留，几棵大树"穿"过屋顶，把树桠伸向天空。水池中经年不息的潺潺流水，室内阳光东升西落洒下的光影，对置身其中的人们产生积极而又和谐的影响，从而激发人们对传统文化的溯源和对设计探索的创新！

锐美集思创意中心是一个与天空对话的设计作品，设计师怀着对一草一木的爱心，在创作中不浪费一片树叶、不浪费一缕阳光，疏影横斜、阳光斑驳、六时皆有境，这自然就是最美的新东方设计！

RUIMEIJISI Innovation Center is located in the GIDC (Guangdong Industrial Design City). It was transformed from an old canteen which provided lunch for over 800 designers. On Dec. 9th, 2012, Xi Jinping, President of RRC and General Secretary of the Communist Party of China, visited GIDC, now there are more than 2000 designers, and the old canteen has become a design innovation center. The designer remodeled the whole building completely and applied a new oriental style in it. This new style combines a simple and fashionable aesthetic trend, integrates tradition and modern, which is an interpretation of China's traditional style, a contemporary design based on the full understanding of Chinese culture, and the inheritance and development of the traditional culture. Trees are maintained , a few big trees cross through the rooftops and stretch into the sky. Flowing water is gurgling in the pool and the sunlight is shining brightly, all these provide a positive and harmonious atmosphere and inspire people to pursue innovation on traditional culture and design.

The designer created the RUIMEIJISI Innovation Center with caring and love, and he took full use of the trees and sunlight. The sunlight filters through the trees and the shadow of trees is reflected on the water, together they establish a dialogue between the Innovation Center and the sky, and this natural scene shows the most beautiful oriental design.

NOMINATION FOR THE BEST CULTURAL SPACE DESIGN

最佳文化空间设计奖提名奖

陈彬（中国武汉）
Chen Bin (Wuhan, China)

获奖项目/Winning Project
Big House当代艺术中心
Big House Contemporary Art Center

设计说明/ Design Illustration

武昌第一纱厂办公楼，建于一个世纪之前的1915年，是民营资本家自筹资金建造的第一幢西式办公楼，隶属于商办汉口第一纺织股份有限公司。由景明洋行设计，汉协盛营造厂建造，当时最好的外资建筑设计公司和最好的建筑公司。与它同时期的厂房已不存在，只有这幢带钟楼的办公楼得以保存下来。

这次，我不仅仅是作为一个设计师参与其中，更是作为一个策划者的身份来面对这次的改造和使用。讨论再讨论，最后达到共识，决定将其改造为一个与当下城市生活方式中，关于艺术和设计有关联的综合形态公共空间。这个曾经在我们对传统生活定义"衣食住行"中与第一位相关的百年老楼，将会被更丰富的当下的生活元素注入而充满生机。

我想应该做点什么来表达这感动，想到了一位墨西哥艺术家的装置艺术，决定借用这种方式！那些纱线，那些色彩！用装置去表述，去记录，这是态度，也是赞美，是对曾在这里绽放而已远去的青春的赞美。

有人说阁楼是灵魂的栖息地。我觉得，每个人心中都有一个属于自己的阁楼。所以，在大楼里看到这间阁楼时，就有了这样一间阁楼客房的设想，因为空间尺度有限，运用服装打版折纸样的方式完成了睡床、工作台、发呆岛、衣柜、音乐沙发、水吧台、洗漱台，同时保留了

原有的老木门和板条墙面，产生一种视觉和触觉的有趣对比。这样的阁楼整幢大楼有两处，所以，这里又变成了只有两套房的艺术酒店，从客房面积与公共配套面积比例上看，应该可以算作超级奢华酒店了！

The office building of Wuchang First Textile Mill was built in 1915. As the first western style workspace funded by private capitalists, it was affiliated with Hankou First Textile Co., Ltd. The office was designed by Hemmings & Berkley and built by Hanxiesheng Construction Company, the best foreign-funded architectural design company and the best construction company at that time. Unlike other workshops created at the same period, this building with a bell tower is the only one that still exists.

This time, the designer also worked as a planner and participated in the modification of the building. After several discussions, the group decided to transform the current building into a comprehensive public space that caters to the contemporary urban lifestyle, and connects to art and design. This old building was once defined with the traditional idea that basic necessities of life are the most important, but it will be filled with vitality through rich contemporary living elements.The designer was so touched that he decided to use an artistic setting by a Mexican artist. The yarns and colors express and memorize his gratitude, and this is a kind of attitude, also a praise to the lost youth. Someone said the attic is a habitat of soul. The designer thought that everybody has an attic within themselves; therefore, he came with an idea of designing an attic room.

Because of the limited space, the designer finished the bed, worktable, wardrobe, sofa, bar table and water basin in a way that is similar to the folding pattern of clothing, meanwhile, he maintained the previous wooden door and lathes, all these form an interesting visual and sensory contrast. There are two attics inside the building, seeing from the ratio of the guest room area and public space, it is like a luxuriously artistic hotel which only has two suites.

202
The Best
Exhibition
Space
Design Award
最佳展示空间设计奖

艾特奖

最佳展示空间设计奖
THE BEST EXHIBITION SPACE
DESIGN AWARD

INTERNATIONAL
SPACE DESIGN
AWARD

获奖者/Winner

梁穗明 （中国广州）

Liang Suiming (Guangzhou, China)

获奖项目/Winning Project

帆·构想/Sail·Vision

获奖项目/Winning Project

帆·构想
Sail·Vision

设计说明/ Design Illustration

PONE ARCHITECTURE设计的《帆·构想》位于武汉黄金中轴光谷大道南片区，周边遍布十几所校园，环境气息充满年轻人的自信与活力，他们勇于接受新鲜事物的特点，激发我们创造一个体验空间——与传统对立，在紧张和动荡的氛围中激活人的思维。

销售中心占地面积690平方米，因为外向型的坡屋顶建筑和主体是被分为前半部分14.7米高，后半部分仅有2.8米高，我们需要打破这个呆板的构造。希望在这些限制性条件中，打碎、叠加、重组设计。面对现有的条件制约，我们充分利用建筑斜面，塑造一个线性解构的空间。空间形态如一个巨大漏斗，吸纳人流，引领参观者进入体验性的展示空间。

室内空间五层叠扣，最外层三角木饰面内扣第二层白色体块，第二层白色体块内扣第三层木色体块……从大到小，一环扣一环，有节奏地层层递进，既相互连接，又相互对立，令这个反差极大的突兀空间能够默契连接，将空间的高度最大化利用，并达到合理分明。大面积整体地面，与设计板块浑然一体，统合纯粹、干净的空间感。空间挑战极致，在内部创造天花零灯孔的极简灯光系统。

中空天花上吊装了一组由71条细金属线和20条带光源粗木管的非对称性三角几何团组装置，线条与面块在空间秩序交错，生成活性节奏。利用地灯及装置的灯光，运用漫反射原理，让第二层白色三角体块作为大堂最重要的一块灯片，通过折射，最终形成空间整体柔和明快的灯光氛围。

中空吊装71条细金属线和20条带光源粗木管，组成线性结构灯光装置。光与线编织交错，与空间面叠加，赋予空间灵动的生命和变动的性格，平衡与失衡构造在此被重新定义，构筑悬浮的"巢"。进入到空间，映入眼帘的是一个由多组三角切面构成的几何形前台和相同形式的沙盘模型展示台。从第三层木色体块进入到内部的洽谈区，整体运用三角切割，区分出几个不同功能区。

这里值得一提的是空间左侧的三个窗洞，运用一个向前倾斜的面，构造出三个斜形四边体，并为其特别定制三组卡座沙发将空间连成一体，既呼应了空间的整体线性设计，又最大限度地合理利用内空间。座位有序、自由地散布在每一个角落，沙发切割人流动线，年轻时尚的色彩配搭线条概念性的软装，激发对未来生活的想象。

"帆·构想"创造一个与传统对立、不安定、紧张、动荡的空间，激活年轻人思维，激发创造力，用全新的思维去构想未来立体生活。

Sail · Vision from PONE ARCHITECTURE, located in the southern regions of Wuhan Optics Valley Road, the golden axis of Wuhan City, with the construction area of 690 square meters, is a building with an outward sloping roof. The main body of the building is divided into 2 parts with huge height difference between the first half of 14.7 meters and the second half of 2.8 meters. We need to change the dull construction by breaking, stacking and restructuring the space. Given the current limitations, the building slope and linear deconstruction theory are fully used to introduce a brand new order. The internal shape of the building is like a giant funnel which attracts visitors to enter and experience the exhibition space.

The indoor five layers overlapping together, the outer triangle wooden block locks the second layer of white blocks, which internally interlocks the third layer of wood-color blocks...From big to small, one layer to another, they rhythmically move forward in-depth, and make the lofty space connect smoothly with huge contrasts, achieving maximum and reasonable utilization of the space height.

The 71 cords of thin metal lines and 20 thick wood tubes with light are hoisting in the medium altitude, to form the lighting equipment of the linear structure. With the weaving of the light and the line overlapping with the space surfaces, the balanced and unbalanced structure is redefined, and a floating "nest" is made. The ceiling challenges to build a minimalist lighting system with zero light hole. The white triangle block becomes the most important piece of light in the lobby. The linear lighting equipment and the light diffuse reflection of the floor lamps create a soft and bright atmosphere. With the linear cutting of the whole space, the reception and the showcase sand table model are formed by multiple sets of triangular sections. From the wooden block of the third layer to the meeting area, the place is separated and arranged by trigonometric tangents for the functional areas.

The three windows on the left are refined by the slopes inclining forward, and the three sets of customized booth sofas connect the spaces as a whole. The soft orange ornaments adorn this pure space, and the bright color scheme totally relaxes the whole internal atmosphere, and inspires people's imagination. This project creates a space which is opposite to the traditions, and it's full of instability, uncertainty and tension, aiming to activate young people to think out of the real-life rules, and to use new logic to imagine a stereo vision of the future life.

SAIL VISION
PONE ARCHITECTURE
WWW.PLZ.HK

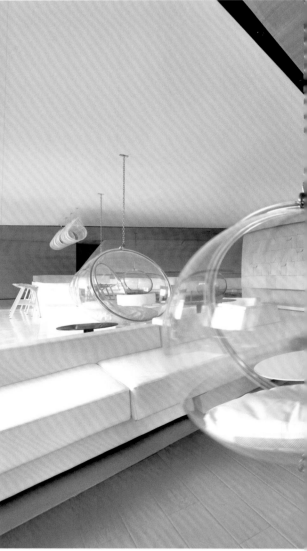

获奖评语

这个展示项目很有趣，它很好地重新界定了室内展示空间。使用一些元素，便重塑出一首空间的诗歌。所有元素条理分明，在最华丽的空间内融为一体。

This is a really interesting project because it has ability to redefine the interior space of this exhibition project. And making this with these few elements is able to redefine all the poetry of this space. All elements are coherent and work inside the most glamorous space.

NOMINATION FOR THE BEST EXHIBITION SPACE DESIGN

最佳展示空间设计奖提名奖

何思玮（中国广州）
He Siwei (Guangzhou, China)

获奖项目/Winning Project

流动晶体
Flowing Crystals

设计说明/ Design Illustration

这是一个没有中空，室内较多梁结构的多边形建筑，作品从火山喷发的壮观场面得到灵感，模拟地壳运动，创造晶体概念，突破空间规整格局，唤醒并激发访客空间的感知与想象。

人们仿佛置身地壳岩层之中，亲身体验金色不锈钢板块与钢琴白烤漆板块在地壳运动的作用力下，产生强大的挤压力量，发生变形，炽热的岩浆冲破岩层的围压，瞬间喷涌而出。立面重重相叠的银白色木条肌理，呈现出独特的光影瀑布，雾化了空间。来往的访客望着墙面，仿佛看到矿物质在流动。天花、墙身、柱体被精心切割，通过光的作用产生折射，让空间散发出璀璨的钻石火彩。原场地的高差，用线条去修葺，规整空间，形成层次错落分明的岛屿式洽谈区域。座位有序、自由地散布在"岛屿"的每一个角落，沙发成为"岛屿"与"岛屿"之间的界限。建筑的内部空间呈现一种原始的野性魅力。

This is a polygonal building which has several beam structures. Inspired by the spectacular scene of volcanic eruption, the designer simulated the crustal movement and created the concept of crystal, which breaks through the regular layout and motivates visitors' perception and imagination of space.

People seem to stay beneath the earth's crust, which allows them to experience how the golden stainless steel plates and white piano plates deform under the strong extrusion caused by crustal movement, and the hot magmas break the rock stratum and spew out immediately. Vertical stacking silver white wood veneers present a unique waterfall of light and shadow, when visitors looking at the wall, they would feel like the minerals are flowing. Ceiling, walls and cylinder were carefully cut, and the refraction of light gives the space dazzling radiance like diamond. The original site was uneven, so the designer applied lines to arrange the space and created a well-proportioned island style discussion area. Ordered seats are freely scattered in every corner of the islands, and the sofas become the boundaries between them. The inside space shows an original wild charm.

9 SALES CENTER
TY CRYSTAL

首售中心

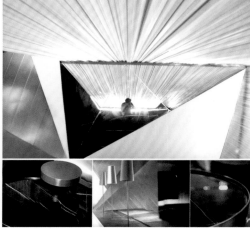

NOMINATION FOR THE BEST EXHIBITION SPACE DESIGN

最佳展示空间设计奖提名奖

Studio TAMassociati
(意大利/Italy)

获奖项目/Winning Project

2016威尼斯建筑双年展意大利馆
Italian Pavilion at the Biennale Architecture 2016

设计说明/ Design Illustration

TAKING CARE（关怀）是2016威尼斯建筑双年展意大利馆策展公司TAMassociati（该公司的合伙人是Massimo Lepore、Raul Pantaleo和Simone Sfriso）选取的关键词，以响应此届威尼斯双年展主题"Reporting from the Front（前线报道）"。策展人说道："关怀是此次意大利馆想表达的一种行为，然后在公众中生根与传播，得以促成一种新公民意识产生。"这是建筑的一种很高的社会视角，也是其根基。主题分为三个部分："思考"、"遇见"和"行动"，在此三个议题的基础上讨论公共利益。它令参观者去辨识身边的类似事件；令机构去推动这个事件，即使只拥有有限资源；令设计师在建筑中表达它。

一种流行语言

当谈论外城时，大部分时候人们的语调都是暗淡和顺从的。但在意大利馆，我们希望表达边缘人的生活：故事、爱、智力、激情……出于这个原因我们选择了一种"流行"语言，通过文笔生动的小说，创造一种特别的情感。小说将故事放到一个不同的地方，主人公肩负将乐观主义的信息传达到未来的任务。

低成本/社会价值

展览的装置成本很低。低成本表示高效率和去除冗余，同时优化服务和成本，这样最终用户和社会将得到便利。低成本设计是一个绝佳机会，去专注于消费的本质并发展出一套不同的哲学，从社会效用的真正参数开始创造价值。这逐渐变得必要，展览尝试去展示一种简化，从而在经济和效率中储存技术和成本。威尼斯建筑双年展是一场短暂的大型活动，呼吁降低浪费是装置的基本主题。意大利馆重新利用之前艺术展的照明设备。展览的设计保存了一些已安装的石膏板分区，再利用了米兰世博会上爱尔兰馆的夹层木板。一些简单的表达，但与展览想传达的对资源的关注保持一致。

Taking Care – designing for the common good curated by TAMassociati

TAKING CARE is the keynote that TAMassociati (Massimo Lepore, Raul Pantaleo, Simone Sfriso) – the curatorial team of the Italian Pavilion at the Biennale Architettura 2016 – has chosen in keeping with the theme of Reporting from the Front proposed by Alejandro Aravena to recount an architecture created with communities and capable of sharing needs and aspirations. "Taking Care", say the curators, "is an action that arises in the Italian Pavilion at the Biennale Architettura 2016 and then takes root and spreads outside it to generate a new civic awareness". It is a highly social vision of architecture, but also the root of its being. The theme is divided into three sections, 'Thinking', 'Meeting' and 'Acting', resting on three specific issues involving the common good. It urges visitors to recognize it, when it is found around them, the institutions to promote it, even with limited resources, and designers to embody it in architecture.

A pop language

When talking about the outer city, all too often the tone is sombre and resigned. But in the Italian Pavilion we want to speak of the life hidden in the peripheries: stories, love, intelligence, enthusiasm… For this reason we have chosen a "Pop" language, which, through the graphic novel, creates a special empathy with the stories we tell. The graphic novel helps relocate the story in a different place, with the main character being given the task of bearing a message of optimism for the future.

Low Cost/Social Value

The exhibition installation is low cost. Today low cost basically means increasing efficiency and eliminating the superfluous, while optimizing services and costs to the benefit of end users and society. Low-cost design can be an extraordinary opportunity to focus on the question of the real nature of consumption and develop a different philosophy of creating value starting from parameters of real social utility. It is increasingly necessary, as the exhibition design seeks to show, to simplify in order to restore

technology and costs to economy and efficiency. In the context of the Biennale, a large temporary event, care to reduce waste is a fundamental theme of the installation. In the Pavilion, the display reuses lighting equipment supplied for a previous art exhibition. The exhibit design retained some plasterboard partitions already installed, it reused the laminated wood panels of the Irish Pavilion at the Milan Expo. A simple gesture, but consistent with what the exhibition aims to show in terms of a concern for resources.

NOMINATION FOR THE BEST EXHIBITION SPACE DESIGN
最佳展示空间设计奖提名奖

邵唯晏（中国台湾）

Shao Wei-Yen (Taiwan, China)

获奖项目/Winning Project

台北公共住宅展
2016 Taipei Public Housing Exhibition

设计说明/ Design Illustration

城市发声·开放性·对话
公民意识抬头，居住正义成为全民关心的议题，透过公共住宅展览创造一个城市与公民交流的平台。以城市建筑量体为空间概念发想，期望以开放思维走访民居生活，体验场所精神，并透过展场再诠释多面向的公共住宅议题。内容囊括城市未来发展的轮廓，用真实史料记录公共住宅发展，探讨法规与制度限制下的局限性，设计开放性的互动经验开启对于未来居住的想象。

白色力量
白色象征着纯洁、和平、正义，同时也是柯市长身穿白袍行医期间济弱扶贫的精神象征。让公共住宅承载这白色精神，发挥最大的居住正义。

漂浮之岛
落实公共住宅是台湾青年共同的愿景，于展场设计上秉持这般信念，采用全玻璃打造一座象征希望的漂浮之岛。岛上承载着公共住宅建筑模型，透过光影投射在地面上，意喻着这些愿景未来都会完整呈现在台北的土地上，公共住宅将是日后住宅结构不可或缺的一部分。

一坪空间
透过一坪大小的空间，以大型量体吸引参观者目光并且引导入展间，而围塑出来的空间象征着住宅空间是人民的基本需求，非奢侈品。

家•巷
在空间中置入家的意象量体，唤起民众对家的渴望与想象，让民众可以品味三维空间与二维平面交迭而成富有层次的生活况味。

The Voice of the Citizens·Openness·Dialogue
As citizen consciousness is rising, the living justice becomes a subject which all the people are cared about. The public housing exhibition creates a platform, through which a city can communicate with its citizens. Inspiring from urban architecture, we expected to understand the spirit of the place by experiencing residents' real life with an open view, and to explain various public housing issues through the exhibition. The contents include outline of the city's future development, historical data which record public housing development, and discussion of limitation under laws and regulations. An openly interactive experience is for visitors to explore future living.

The White Power
White symbolizes purity, peace and justice. Meanwhile, it is the spiritual symbol of the Mayor Ko Wen-je during the time when he was wearing a white robe helping the patient. Letting the public housing carries this spirit is to maximize the living justice.

The Floating Island
To achieve the goal of public housing is a common vision of Taiwan youths. With this belief in mind, we used glasses to create a floating island which symbolizes hope. The island bears public housing models, which are projected on the ground through the light and shadow, implying that the vision will be fully realized in Taipei, and public housing would be an integral part of the future residential structure.

3.24 m2
Through a 3.24 square meter space, the big cube attracts visitors and guides them into the exhibition stand, and the framed space is telling visitors that housing is a basic need of people, not a luxury.

Home·Alley
The house's image frame arouses people's desire and imagination of home, allows them to enjoy a multilayered life which is formed by three-dimensional space and second-dimensional plane.

五大願景方案
Beijing Projects in Prospect

空中看台北
Taipei from the Air

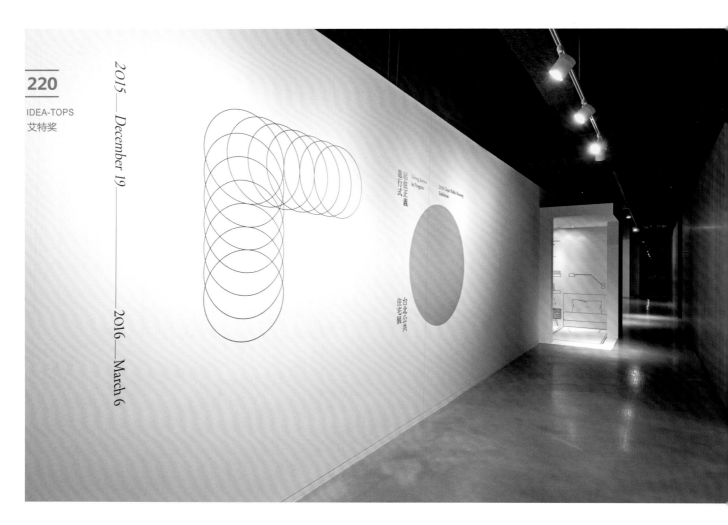

2015——December 19

2016——March 6

居住正義
進行式

台北公共
住宅展

Living Justice
in Progress

2016 Taipei Public Housing
Exhibition

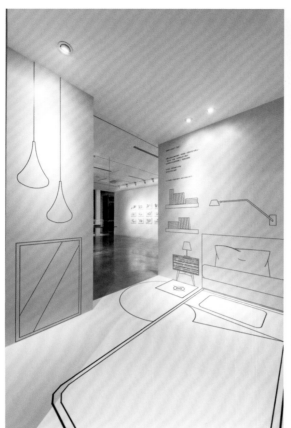

NOMINATION FOR THE BEST EXHIBITION SPACE DESIGN

最佳展示空间设计奖提名奖

尺道设计事务所（中国佛山）
C.DD (Foshan, China)

获奖项目/Winning Project

回 | 万物归宗
Hui | Origin of Everything

设计说明/ Design Illustration

这是一个承载故乡感情及对万事万物追本溯源思考的建筑装置。

3米×3米×3米的立方体，自上而下俯瞰整个空间的横截面，竹子墙和外立面墙的两个方形拼出了一个象征"归宗"的"回"字。

一条能容纳一人通行的单向走道就此辟出，观众从外面的世界由狭小的入口独身进入，和着带有神秘的背景音乐及虚幻的光影缓步绕行一周，透过根根竹子的间隙，看到摆在空间正中央的南狮和牛皮鼓白胚，犹如经历了一次环境和内心都由外到内的旅行。

从空间层次上看，从外界环境，到外立面的钢板材料，到内墙的竹子，再到被竹子围起的传统原型，还经历了由现代到古老，由新到旧的过程。

装置以奇妙的方式邀请参观者穿越游历，旨在讲述人类在探索科技与追根溯源之间，重拾生命至高本源的事实。

This is an architectural installation containing the emotion to the hometown and the thinking to trace to the source of our modern society.

It is a cube of 3m*3m*3m, when looking at the cross section of the whole space from above, the two squares created by bamboo wall and exterior wall together form the Chinese character "Hui" (回), which symbolizes "return to the origin".

A one-way path which allows passage of only one person is built, from which visitors could come into this space alone through a narrow entrance. With the company of mysterious background music and illusive light and shadow, while visitors walking slowly around the space, they can see a Chinese southern lion and a Chinese leather drum by looking through the gaps between the bamboos; it is like a journey from outside to inside, both of your inner self and the environment.

In terms of spatial level, from the external environment, the steel

plate material of the façade, to the bamboo of the inner wall, and then
to the traditional prototype which enclosed by bamboos, all allow
visitors to experience the process from modern to ancient and from
new to old.
This installation invites visitors to travel in a wonderful way. It aims at
expressing the fact that human try to regain the most primitive origin
of life during the process of exploring technology and tracing back to
the source of everything.

ortation
Design Award
通空间设计奖

艾特奖

最佳交通空间设计奖
THE BEST TRANSPORTATION
SPACE DESIGN AWARD

INTERNATIONAL
SPACE DESIGN
AWARD

获奖者/Winner
Zechner & Zechner
(奥地利/Austria)
获奖项目/Winning Project
格拉茨总站/Graz Main Station

获奖项目/Winning Project

格拉茨总站
Graz Main Station

设计说明/ Design Illustration

格拉茨中心车站，每天接待30000名旅客，是奥地利最重要的交通中心之一，在未来的重要性会更高。当地铁路系统完工后，在高峰时候将以15分钟为周期，每日给格拉茨带去40000名旅客，旅客来自全国各地。这表示格拉茨中心站的设施要达到预期载客要求，同时拥有旅客数量增加的空间。这要求大量的建设工作，这是格拉茨中心车站的任务之一。

格拉茨总站的新顶棚——"波"
格拉茨中心车站的站台顶棚不再适应当前的基础设施建设要求。顶棚被改造过好几次，有些改造只是暂时之计。此外，部分重新配置的站台要求顶棚被拆除。建筑师利用这个机会，设计了一个新顶棚。这个新顶棚将超越当前站台区域，对已经修复并扩建的接待大楼以及未来计划的车站广场顶棚来说都十分具有吸引力。梁的两侧跨度各超过40米，跨越通道楼梯之间的区域，形成一个"波"。减少柱的数量以及距离顶棚8米的空间为在中心区域候车的乘客提供了宽敞的遮盖。横跨的梁有4米高，在肩部位置相互对称。梁之间的部分向上突出形成一个V形，产生一个沟壑，为站台提供照明，并且被半透明的膜顶棚覆盖。顶棚的两个横梁跨越了整个轨道区域，为中心站台区域提供了宽敞的顶棚。
顶棚结构在中心联结，在站台外区域形成独立的舌状站台顶棚。这创造了一种像波一样的总体形象，在延展过程中逐渐消散。"波"的形象作为一个象征，强调了火车这种交通方式的活力。

150米的艺术
车站里北部是一个新建成的人行通道。新通道可以通向所有站台，同时也可到达轨道的另一侧。通道很长，但这种很长的感觉通过"创造宽度"、优良的人造照明所减弱，以及最重要的是通道使用了Peter Kogler设计的一个150米长的艺术作品。这个作品采用玻璃喷镀的形式，进行大面积喷镀，将通道变成一个非常特别的展览空间。艺术墙是车站大堂里装置的一个深化，这个装置开始于2003年，作者也是Peter Kogler。

Graz central station, with 30,000 travellers per day, is one of the most important transport hubs in Austria, and its importance will be increasing further in the future.Completion of the local rail system will, at peak times, bring 40,000 passengers per day to Graz on a 15 minute cycle, from every direction. This means that the facility requires adjustments to cope with expected current and future service requirements

and increase in capacity. This will require extensive building work, which is combined under the Graz Central Station project.

The New Roof at Graz Station – the "Wave"
The platform roofs at Graz central station no longer met current infrastructure building requirements. The roofs come from several different periods, and some were constructed as temporary measures. Additionally, the partial reconfiguring of the platforms requires that the roofs be demolished.
The architects made use of this opportunity to build a new roof that will go beyond the area of the current island platform roof and form an appealing counterpart to the already restored and extended reception building and the future projecting roof in the station plaza.Beams with two spans of over 40m each, arch over the area between the tunnel stairs, forming a "wave". The reduced number of supports and a height of up to 8m

below the roof create a generous cover above the central area for the waiting passengers.The arched beams, which reach a height of 4m at the shoulders are situated in pairs. The distance between the beams increases upwards to form a V, producing a lens shaped gap between the pairs of beams that will provide illumination to the platforms and is covered by a translucent membrane roof. The crossbeams of the roof structure run across the entire track area to provide a wide roof in the central platform area.The roof structure, which is connected at the centre, transitions into individual, tongue shaped platform roofs in the outer areas of the platforms. This creates an overall impression that resembles a wave gradually dissipating as it spreads. The image of the "wave" acts as a symbol that emphasizes the dynamic of the train as a mode of transport.

150 Meters of Art

In the north of the station site a new pedestrian tunnel was constructed. The new passage provides access to all platforms, but also provides a connection to the area of the city on the other side of the tracks. The impression of length of the tunnel is reduced by creating width, good artificial lighting, and, most importantly, by installing a 150m long artwork by Peter Kogler, in the form of glass cladding with large printed areas, making the passage a very special exhibition space. The art wall is a continuation of the installation in the station hall, begun in 2003, also by Peter Kogler.

获奖评语

这个设计
把交通和城市空间的
创造结合在一起
形成城市
值得记忆的标志。

This design
combines the traffic
with urban space
creation to
create a city's
landmark that is
worth remembering.

NOMINATION FOR THE BEST TRANSPORTATION SPACE DESIGN

最佳交通空间设计奖提名奖

MXT Studio (Maximina Almeida, Telmo Cruz)

(葡萄牙/Portugal)

获奖项目/Winning Project

葡萄牙人行与单车桥
Pedestrian and Cycling Bridge, Portugal

设计说明/ Design Illustration

城市线路在发展过程中不断重叠和聚集，这些都会呈现在地图之上，它们见证着延续和断裂、欲望和价值，过去、现在和未来。这座天桥一定会出现在未来地图中，会变成另外一个旅游地区及交通方式，因为它融入了一种全新的全球生态理念。从2009年到2014年，这个国际化的桥梁设计项目从竞标到开始设计，期间里斯本的自行车车道从40千米增加到80千米，见证了这个城市自行车数量的剧增。

为了鼓励将休闲活动和骑自行车联系在一起，桥架一开始被安置在北部城市的高地上，高度约100米，旁边是相对较新的居民区；之后被安置在一块临近滨水的区域，高度为5米。两个场地都是绿色区域，有合适的空间建设远离汽车道路的自行车车道。现阶段的目标是为推广使用自行车创造条件，推动将自行车变成另外一种无污染的交通方式，并希望能够扩展至整个城市。这座桥在这个城市的自行车网络中并不那么突出，但它把建筑和休闲绿色空间结合在一起，并与汽车竞争空间。在建设场地中，本菲卡、卢兹和泰尔艾拉农场的车道、林道都将形成一

个新的地图版块。新的通道拔地而起，部分区域让人回想起旧的农场车道，整座桥横跨公路主干道，不同尺寸、形态各异的线条在空间重叠交错。从竞征方案到最终的建设方案，项目不断得到完善，在建造、解决方案和技术要求的每个方面都力求完美。建成之后的桥是一个混合结构，中间没有柱子，但依靠一个具有更大跨距的箱梁结构支撑，这个结构约42米，由8毫米和12毫米厚的钢板组成，而钢筋混凝土箱梁的墙面为250毫米，所有坡道和楼梯的跨度约25米。钢面板和混凝土桥面被内嵌于一个3.3米的等边三角形内，巧妙的建筑细节让各个区域相互连接，使得结构和造型都非常连贯。

在完善各种解决方案的过程中，设计团队谨慎地将成本控制在合理的范围之内，比全钢的设计成本低了近35%，让桥的建造具有可行性。在桥的维护和照明方面，同样经过了不断改善，以求优化技术成本和建设成本。扶手和护栏系统化地运用了三角结构，增强了牢固性，减少了固件和钢材截面。此外，将LED技术运用到护栏的顶部区域，减少了操作成本和维护成本。许多制造商在这一方面已经有现成的解决方案，但该设计公司却在有限的预算内设计出一种能够在公共空间中满足特定技术和安全需求的新的照明装置。

In cities all moments of their evolution overlap and accumulate, multiplying into distinct maps, witnessing its history of continuities and discontinuities, desires and values, past, present and future. This bridge belongs to the maps of the future, inevitably optimistic, of other journeys and transport modes, soft, resulting from a new global ecological awareness. From 2009 to 2014, the time that has elapsed between the international architectural competition for the design of this bridge and its opening, the extension of segregated bike lanes in the city of Lisbon increased from 40 to 80km, witnessing the growing presence of bicycles in the city. Strategically planned to promote the use of bicycles associated with leisure activities, they were initially implemented in the northern city plateau with an elevation of circa 100m, occupied by relatively newly built neighborhoods, and more recently along the waterfront with the approximated elevation of 5m, both locations with green spaces of reasonable dimensions available for the integration of bike corridors, segregated from automobile routes. The current phase of the promotion of daily bicycle usage aims at creating conditions for integrating and promoting the use of bicycles as an alternative means of transport, non-polluting, and, hopefully, extended across the entire city territory. This bridge is one of the first moments where the cycling network of the city, leaves a condition of relative invisibility, by virtue of the implementation strategy initially associated with leisure and green spaces, competing directly with the automobile for space and legitimacy in the city. At the place and context of this bridge project, lanes and tree alleys, of the Benfica, Luz and Telheiras farms, are an opportunity to build a new map. A new network of paths literally taking off the ground, partially a memory of the old farm lanes, cross over 2a Circular, a major arterial road, in a song of overlapped lines of different scales and modes. From the solution proposed in the competition to the built solution the project development process went through the optimization and perfecting of all elements of the construction,

integrating solutions and technical requirements. The built solution is a hybrid structure, without pillars under the central span, resorting to box girders in the larger spans over 2a Circular, about 42m, made of steel plates with 8 and 12mm thickness, and reinforced concrete box girders, with a wall thickness of 250mm, on all ramps and access stairs where the bigger spans are around 25m. Both, the steel and concrete deck, are fully inscribed in an equilateral triangle with 3.30m sides, and their connection was carried out using an ingenious structural detail to keep structural and shape continuity.

The deliberate attention to the optimization of all solutions maintained costs within acceptable limits, approximately 35% below an all–steel solution, thus making feasible the bridge construction. In what regards the protection and lighting elements, these were also perfected in order to optimize both the technical solution and construction costs. In the case of handrails and guardrails the systematic use of triangulated supports helped increase its strength, reduce the number of fixings and the section of the steel profiles. Integrated into the guardrail top profile the lighting was carried out using LED technology thus lowering operating and maintenance costs. In the absence of a solution already available in the catalogs of many manufacturers the design firm chose to develop, together with an international manufacturer, one light fixture capable of fulfilling, within the budgetary limits, with the specific technical and safety requirements of the installation of lighting in a public space.

NOMINATION FOR THE BEST TRANSPORTATION SPACE DESIGN

最佳交通空间设计奖提名奖

GOTTLIEB PALUDAN ARCHITECTS

Gottlieb Paludan Architects
(丹麦/Denmark)

获奖项目/Winning Project

欧登塞Byens Bro人行与单车桥
Byens Bro, Foot and Cycle Bridge, Odense

设计说明/ Design Illustration

这座桥叫Byens Bro（城市之桥），联结了欧登塞中心区和新城市发展港口，也是通往铁路站台的更好的通道。这座桥创造了丹麦第三大城市欧登塞的新地标。
Gottlieb Paludan建筑事务所担任项目设计师和主咨询师的职责，其他参与的咨询公司有ES-Consult, NIRAS, Bartenbach和照明设计艺术家Anita Jørgensen。项目施工单位是Bladt Industries A/S。桥梁在2015年夏季开放。
新的人行和单车桥为欧登塞市政厅提供了一个构想：单车和行人如何通过简单高效的建筑工具穿过铁路。
这座桥是欧登塞城市空间的一个多样化的和壮观的延展，因为它联结了铁路站台的南北两侧。这成为欧登塞城市空间美丽、实用和有机的一部分，所有行人、单车和铁路站台出来的人每天都可以方便地使用这座桥。桥梁的功能使得建筑和工程艺术的界限变得模糊，赋予欧登塞一个启发性的和高效的交通解决方案，同时创造了一个恒久存在的地标。
这座桥被设计成两座桥合二为一：一座单车桥让骑单车的人们通过，一座人行桥让人们步行。

出于舒适性和安全原因，桥的这两个功能被分开，但在视觉上和建筑上，它们被合并为一座桥。该桥梁获得了2015年欧洲钢结构设计奖和欧登塞"美丽建筑和结构"奖。

The bridge, named Byens Bro (The City's Bridge), provides a link between Central Odense and the new urban developments by the harbour, together with better access to the platforms of the railway station, while it at the same time creates a magnificent new landmark for Odense, Denmark's third-largest city.
Gottlieb Paludan Architects has acted as design architect and lead consultant on the project with sub-consultants ES-Consult, NIRAS, Bartenbach and lighting-design artist Anita Jørgensen. The construction project has been carried out by Bladt Industries A/S. The bridge opened in the summer of 2015.
The new foot and cycle bridge gives shape to Odense Municipality's vision of a transport link for cyclists and pedestrians across the railway, using simple and effective architectural tools.
The bridge is a dynamic and spectacular extension of the existing urban space of Odense, as it connects the urban quarters north and south of the railway. It has become a beautiful, practical and organic part of Odense's urban space, and easy to use for all pedestrians and cyclists and for the many people travelling to and from Odense by train every day. This basic functional quality translates into a bridge that blurs the boundaries between architecture and engineering artistry and gives Odense an inspirational and efficient traffic solution as well as a striking landmark that will stand the test of time.
The bridge is devised as an extension of Odense's existing infrastructural grid and therefore it is actually two bridges in one: a cycle bridge designed to cater for cyclists' traffic needs and a foot bridge providing the very best access conditions for pedestrians. For comfort and safety reasons, the two functions are segregated but visually and architecturally, they are brought together in one bridge. The bridge has received a European Steel Design Award of Merit in 2015 and a prize for 'beautiful buildings and structures' from the city of Odense.

NOMINATION FOR THE BEST TRANSPORTATION SPACE DESIGN

最佳交通空间设计奖提名奖

Rintala Eggertsson Architects
(挪威/Norway)

获奖项目/Winning Project

Tintra人行桥
Tintra Footbridge

设计说明/ Design Illustration

设计

新桥梁的设计希望将桥梁设计成有节奏的一个个模块，使过河的时间变为一个个间隔，以减少过河时的单调感。桥梁设计以旧桥遗留下的河两边的两根柱子之间的距离为基础。

主要的钢结构形成主节奏，木结构形成副节奏，就像打击乐器一样。两个柱子上的桥廊给节奏制造了停顿，否则旋律略显寻常。桥廊也给过河这个线性活动提供了休息的间隙，也能和景观更直接地接触。它们是钓鱼的理想场所，彼此之间也不会互相干扰。

在洪水中倒塌的旧桥是悬索桥。新桥我们想设计得不一样。由于桥的两种原材料，钢和木材，自然地我们将两种材质分开，一个在里一个在外。新桥外面采用了钢结构用来支撑桥的整体重量，里面则采用了木结构给人们一种宜人的观感。

材料

就桥梁而言，一般客户最关注的就是维护费用。因此他们对使用耐候钢作为主结构以及固雅木为包层材料很感兴趣。这将结

合强度和耐久性，并且最重要的是，多年都无需维护费用。客户也喜欢桥的顶棚设计，这可以挡雪，而扫雪工程对人行桥来说很复杂。

DESIGN

The design of the new bridge was generated by the idea of a rhythmical play of modules crossing the river in order to break down the timeline into intervals and reduce the monotony when crossing the river. This play of modules would be based on the distance between the already existing pillars in the river, the only remains of the old bridge.

The main steel structure would form the base rhythm and the wooden structure a secondary movement, similar to what you would find in a score for percussion instruments. The two rooms on top of the pillars were therefore used to generate pauses in this otherwise regular rhythm, offering a break from the linear movement across and a possibility of more direct contact with the landscape. They were also ideal as fishing platforms as the two activities would not need to disturb each other anymore.

The old bridge which collapsed in the river flooding was a suspension bridge with a clear visual language. That clarity was an important point of departure for us in the design of the new bridge. But given the two main materials, steel and wood, it became natural to differentiate between the inside and the outside and offer the two materials different roles so to play. The steel would express the raw forces going through the bearing structure on the outside and the other would communicate sensorial information on a human scale on the inside.

MATERIALS

As it often is the case with bridges, maintenance was an important issue for the client. They were therefore very intrigued by the idea of using Cor-ten steel as a main structure and Accoya as a cladding material. That would combine strength, durability and above all no costs of maintenance over the years. They also liked the idea of covering the bridge with a roof, which would eliminate the need of snow plowing, which can be complicated issue for pedestrian bridges.

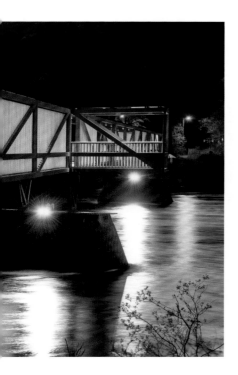

NOMINATION FOR THE BEST TRANSPORTATION SPACE DESIGN

最佳交通空间设计奖提名奖

GaP Architectes (法国/France)

获奖项目/Winning Project

格勒诺布尔停车楼
Parking Building in Grenoble

设计说明/ Design Illustration

这座五层楼的停车场位于Arlequin画廊外，在格勒诺布尔市维伦纽夫区。这是该区域城市更新的第一批建筑。
项目目标是建成一座明亮的楼。光的应用是为塑造清晰的公共空间。
停车场的活动，车辆和人的移动，可以从街上清楚看到，从而增添街区的活力。
由于平面图的简洁，项目是开放且灵活的。首层的部分区域为活动区。
中心的两条螺旋坡道，是一种可塑性表达。坡道可直接通往各楼层，而不必经过前一楼层，没有交叉的交通流量。坡道交通是独立的，可以设置程序化的停车托盘。将中心空间分区，可以举办活动或者设置办公室，而不会影响车辆的停放。

This 5 stories car park is located just outside of the Arlequin gallery, in the Villeneuve neighborhood in Grenoble. It is the first new equipment for the urban renewal of the area.
Our project aims to build a bright building, light, whose function is legible from the public space.

The car park activity, the movement of vehicles and people, are visible from the street and participate in the liveliness of the district.
The project, due to the simplicity of his plan is open and flexible. Part of the ground floor is occupied by activities. The central double helix ramp (type Chambord) is plastically expressive. It can serve each level without having to go through the previous stage and without crossing traffic flows. Traffic in the ramp being independent, programmatic mutation parking trays is possible. By partitioning the central volume, activities or offices could be built on some levels while maintaining the rest of the parking operation.

艾特奖

最佳办公空间设计奖
THE BEST OFFICE
DESIGN AWARD

IDEA TOPS

INTERNATIONAL
SPACE DESIGN
AWARD

获奖者/Winner

M+M Creative Studio

(美国/US)

获奖项目/Winning Project

Edmunds.com网站公司总部大楼

Edmunds.com's Headquarters

获奖项目/Winning Project

Edmunds.com网站
公司总部大楼
Edmunds.com's Headquarters

设计说明/ Design Illustration

"聚集只是开始，相处是进步，共同工作才是成功。"——亨利福特

客户的初衷十分直接，这个位处"硅谷沙滩"的自动化信息公司的CEO想要一个新的独出心裁的空间，首先可以容纳他600多人的团队，并且可以为团队提供合作、打造团队、创造、定义、做白日梦、招聘、讨论、聚焦以及最终共同工作的地方，无论何时都让人们能以最佳状态处理工作。两层楼一共13471平方米，我们在此基础上开展工作。一开始我们进行大的变更：首先，我们改造了不规范的中庭外围，为两层楼创造了一个"高速公路"循环；接着，我们聚焦于一楼的大型会议空间——根据项目设计元素的"主街道"设置主入口、接待处和"毂盖"（公司内唯一一处咖啡吧台）；第三，CEO认为员工应该拥有自己独立的办公桌（"停车场"），因此在二楼我们设计了一个板桌系统，每个人都有自己灵活的办公空间，还有一些小型会议室和一些其他用途的房间，比如安静室（"休息站"）和IT服务台（"加油站"）散落在楼层各处。
此外，929平方米的中心外庭整体被重新设计成一个新的充满活力的室外空间，包括大型社交和私密空间——通过超过39.6米的可撤回玻璃门与邻近的室内空间连接起来。在中庭内，15米长的多重线性植被（薰衣草、茉莉花和袋鼠花）和潺潺流动的水流全都在平行线条上运行（就像看在一条超大尺寸的高速公路上行驶的汽车），在内置LED灯的条形照明下，就像夜里的一条高速公路，车辆在上面行驶，同时以不同形式变换着颜色。

"Coming together is a beginning, staying together is progress and working together is success" – Henry Ford

Our client's brief was straight-forward … the CEO of this "Silicon Beach" automotive information company wanted a newly inventive space, focusing on his 600+ person team first and foremost … a space where they could collaborate, team, create, define, daydream, recruit, discuss, focus and ultimately work however and whenever they felt best fit their current assignments. Working with two floors totaling 145,000 SF we began our work. We started with the larger moves; first, we harnessed the perimeter of the existing rambling internal atrium as the circulation 'expressway' for both floors, second, we concentrated the larger meeting spaces throughout the 1st floor - anchored by a 'main street' of program/design elements - main entry, reception and 'the hub' (i.e. 'hubcap'), their one and only coffee bar for their entire company) and third, because it was critical to the CEO for each team member to have their own individual desk (i.e. 'parking spot'), the bulk of the 2nd floor is comprised of a plank-desking system that allows flex in linear feet per person where required, with additional smaller meeting rooms dispersed throughout and miscellaneous programmatic requirements such as quiet rooms ('rest stops') and the IT service desk (the 'pit stop').

In addition, the 10,000 SF exterior central atrium was completely re-envisioned - a newly dynamic outdoor space with both larger social and more intimate spaces - connected to the adjacent interior spaces with over 130 LF of retractable glass doors. Inside the atrium itself, the multiple 15' long raised linear planters (planted with lavender, jasmine and kangaroo paws) and gurgling water features - all run in parallel lanes (like cars seen from above on a super-sized highway), weaving among the embedded LED strip lighting that, like a highway at night, race and change color with a varying choreography.

获奖评语

很有未来感，同时用了很多工业设计的手法。提供了不同的体验空间，这是个非常好的新的办公空间的范式。

Full of future sense, while using a lot of industrial design practices. Providing a different experiencing space, this is a very good and new office space paradigm.

NOMINATION FOR THE BEST OFFICE DESIGN

最佳办公空间设计奖提名奖

Setter Architects
(以色列/Israel)

获奖项目/Winning Project

以色列特拉维夫Facebook 办公楼

Facebook Tel-Aviv Offices

设计说明/ Design Illustration

Setter建筑事务所设计了Facebook以色列分公司。公司位于特拉维夫市罗思柴尔德大道上的一栋高层建筑内，一共四层楼。

Setter事务所的价值观是空间服务于人，相应地，项目的设计是功能性的、有机的、自然的、真实的，甚至在某些方面十分简洁。

Facebook公司网络体系的精神鼓励员工之间的协作。开放性、流动性、多样化的空间可以保证多种座位和工作安排，独特的不同的交会，透明性和变化，这些都最大程度地保证了员工的灵活性，从而激发自发性和创造力。

可以清楚地看到，建筑的设计强调了一种与周边城市环境的联系。公司位于当地的文化和社交产地，融合了中东、特拉维夫和雅法的元素，以及再生材料的使用。相应地，被选择的材料和建筑元素被重新诠释，产生一些独特的艺术品：比如圣殿时期的百叶窗被改造成用餐区域的复合百叶窗，著名的装饰瓷砖被重新诠释为在水泥地上的印迹花纹，都在自助餐厅区域。

上面提到的元素为用餐区提供了一种当地的感觉。在另一区域，古老的木板被拆开，在中心开放区域重建了一个图书馆。

其他原来的物品包括老旧的太阳能热水器和气管，都被转化成丰富和装饰空间的花架。主导空间设计的另一原则是将户外感带入室内。一面垂直的绿墙被设置在楼层内，它既是分隔墙，也制造新鲜空气从而给经过的人带去一份生活气息。

空间的一个特色元素是楼梯井。楼梯井由钢和金属网围栏组成，金属网在过去的特拉维夫市建造中十分流行。楼梯井的墙上是客户定制的以色列艺术家FALUJA的作品，作品贯穿了整四层楼。作品融入丰富多彩的阿拉伯图形和各种木制元素，创造了一种和谐的三维视觉。

在现代，员工喜欢一种开放的、采光好的和多功能的工作场所，设计的主要目标是使建筑服务于员工并保证多样化的座位和工作区域。地方艺术的物质性方面被重新诠释和恢复为设计元素，将色彩和纹理融合在一起，捕捉眼球且温暖人心。

Setter Architects Designed the Israeli offices of Facebook; Spread over four floors in a high rise building located at Rothschild Boulevard, Tel-Aviv. Synched with the company's values the space serves its inhabitants. Accordingly, the design is functional, organic, natural, real and even simplistic in some ways.

As a part of Facebooks' networking mentality the company encourages collaborative work between employees; Open, fluid, and dynamic spaces that allow for various seating and working arrangements, unique and different encounters, transparency and change, all in an effort to enable the employees maximum flexibility that spurs spontaneity and creativity.

It is plain to see, that there is an emphasis on relating to the urban environment in which the office is located, a local cultural and social habitat that incorporates elements from the middle-eastern, Tel-Aviv and Jaffa principles, as well as the use of recycled materials. Respectively, the materials and architectural elements that were chosen were reinterpreted, yielding unique artifacts such as: window shutters from the settling Templar period converted to a shutter cladded kitchen island, the well-known ornamental tiles reintroduces as imprinted arabesques on the concrete floor, both residing in the cafeteria.

The above mentioned elements provide the restaurant a local urban feel. In another area, vintage wood panels were taken apart and reconstructed to form a library in the center of the open space. Additional local objects include old solar water heaters and gas tanks that were transformed into planters that enrich and embellish the space. Another principle that dominates the space is bringing in the outdoors. A vertical green wall is positioned on one of the floors; it serves as a dividing partition as well as a breath of fresh air exhuming livelihood to all passerby.

One of the feature components in the space is the stairwell. The staircase is composed of steel and the rail of metal mesh net, a material that was very popular in Tel-Aviv's past construction. The wall of the stair shaft entails in it (throughout the four floors) a custom made artwork by Israeli artist FALUJA. The art integrates colorful arabesque graphics and various wooden elements that create a harmonious three dimensional visual.

Nowadays the employees enjoy an open, well-lit and functional workspace; its main goal is to serve the personnel and grant them diverse seating and working areas. The materialistic aspect was inserted into local artifacts that were reinterpreted and reinstated as elements of design, combining colors and textures that capture the eye and warm the heart.

NOMINATION FOR THE BEST OFFICE DESIGN

最佳办公空间设计奖提名奖

成都沃克豪森设计工作室（中国成都）
Chengdu Wokehaosen Design Studio
(Chengdu, China)

获奖项目/Winning Project

CCTV成都宙斯传媒公司办公楼
CCTV Chengdu Zeus Media Company Office Building

设计说明/ Design Illustration

CCTV宙斯传媒作为CCTV央视微电影青春频道旗下的一家专业从事新媒体传播公司，对于办公室的设计诉求，如何用最低的成本体现功能设计上的高效、开放、互动、交流的工作模式，同时在设计风格中把握年轻人追逐时尚与潮流的特点，是本案希望极力表达的重点，也是设计师面临最大的挑战。
整个空间都是白色为基调，干净简洁，接待厅地面酷似裂纹的图案又给空间增加了前卫时尚感，用简洁的钢结构几何构架叠加穿插配以工业风的吊灯使空间显得简洁有序。
半开放的办公空间，透明玻璃与外界相联系，相互连接又互不打扰。抓住眼球的是浅绿色的沙发和波普风的图案，给冷峻的空间增添了一抹色彩。
公共区域（接待厅、走道、楼梯间）采用色温4000K的筒灯点状均布作为背景基础照明，配以3000K色温的射灯作为局部装饰照明以烘托气氛；办公区域采用三基色LED高显色指数的工作照明，照度值为400lex。
主要装修材料：麦兰德艺术涂料、费诺娜水泥砖、喷绘定制墙布、不锈钢板橡木、工业风复合木地板、水泥地坪漆等。

As a professional new media company under CCTV micro movie Youth channel, for the design requirement of building the company, how to use the fewest budget to express the work mode of high efficiency, openness and interaction of functional design, at the same time grasping the chasing fashion feature of young people. All above is the key point that this proposal would like to express and the biggest challenge that the designer faces.

The key tone of the whole space is white, clean and simple. Reception hall ground which is similar to the crack pattern increases a kind of fashion sense to the space. Simple steel, geometric structure and industrial ceiling lamps make the space concise and orderly.

There is half opened office space, and transparent glass connects with the outside without disturbing each other. What catches eyes is the light green sofa and pop art design, which adds a touch of color to the cold space.

Public areas (reception hall, hallways and staircases) are averagely distributed like dots with tube lamps of 4000K color temperature as basic background lighting, with spotlights of 3000K color temperature as local decoration lighting to make the atmosphere more comfortable. The office area uses 400LEX working illumination of three colors LED with high color index.

Mylands art paint, Florina cement brick, printing custom wallpaper, oak with stainless steel plate, industrial composite wood floor, cement floor paint and so on.

宙斯传媒　ZEOUS MEDIA
INTERPRETATION CENTER
TRAINING CENTER

NOMINATION FOR THE BEST OFFICE DESIGN

最佳办公空间设计奖提名奖

严宏飞（中国杭州）
Yan Hongfei (Hangzhou, China)

获奖项目/Winning Project

万境设计事务所
WJ Design

设计说明/ Design Illustration

万境设计位于杭州西湖边上老和山北一座老办公用房的底层。占地面积400平方米，该建筑的外立面条件杂乱，但周边植被茂盛，四季景色分明。

设计师将玻璃、水泥、混凝土融合，巧妙地运用到空间的设计，靠窗部分全部挑空，接大面积落地玻璃采光极好。重做大部分水泥结构，量身定制一些到顶钢木结构家具、装饰架灯，突出了强大的气场，同时保持了场地粗犷工业感。

设计师希望每一位办公室的员工能在设计的浸染中生生相息，焕发出时代、年轻的活力。

整座事务所没有过多的色调，用平面基础点缀软装润色，加上部分装置设计，让场地协调有设计感，混凝土和灯管线直接裸露也成了最好的装饰点缀。似空间本身与设计进行谈话，从而使得办公室内表面和所有组件连接通过最直接的形式呈现于此。置身于其中，就仿佛站在一个地方，视线可以穿透任何地方。这样的环境，也能更好地体验最原始的界面和自然、季节、环境以及空气。

Studio of WJ Design is located in the first floor of an old office building in North of Laohe Mountain nearby the West Lake of Hangzhou. It covers an area of 400 square meters, although the exteriors look disorderly, it's surrounded by flourishing plants and different sceneries in four seasons.

The designer blended glasses, cements and concretes together and cleverly used them into the space. The large full height glasses ensure a good lighting. He reformed the cement structure and added some customized furniture and decorative lamps which are structured by steel and wood, creating a gorgeous atmosphere, meanwhile, maintaining a rough industrial feel.

The designer wanted that each staff can be energetic to show young vitality in this office.

Only few colors were applied in the studio to embellish its interiors, and the designer also added some settings, all these achieve a harmonious space which has a good taste. The exposed concretes and light pipes contribute to the decoration, just like a dialogue between the space and design, the surfaces and all the other components are presented in a most directly way. When entering, it's like you are standing in a place where you can see everywhere. In such an environment, you can experience the most original nature, season, surroundings and air.

NOMINATION FOR THE BEST OFFICE DESIGN
最佳办公空间设计奖提名奖

林经锐（中国杭州）
Lin Jingrui (Hangzhou, China)

获奖项目/Winning Project
上海创意者之家麻绳设计办公室
Waving Rope Office

设计说明/ Design Illustration

业主诉求
场地位于上海浅水湾文化艺术中心内，是一个室内面积只有170平米的办公毛坯房。业主希望在较低的预算下将这个小型办公场地塑造为具有创新特色、能吸引微小型设计创业团队和创业者的众创办公环境。

设计立意
关键词：分享、低碳、通透性、材料建构
1. 塑造创意者之家空间布局——共享、交流、创意。
主体办公空间为开放式分享空间，并以两个回字形流线为结构，实现了空间布局的流动性。平面功能布局根据流动办公的实际需求，流动办公与会议办公错峰叠加于中央长桌办公区，实现空间使用的高效性。因地制宜，将原本公共楼梯通道上无法使用的尽端空间转化为集图书阅览、台阶教室及公司文化展示的多功能复合空间。
2. 塑造低碳环保的办公场所——麻绳、槽钢、原木。
室内材料以自然生态、可回收利用的麻绳、槽钢、实木为主体，降低造价的同时并彰显麻绳作为自然材料的可能性与材质美感；摒弃无用的装饰，还原建材的本真质感，温润的原木、粗狂的麻绳以及带有强烈工业属性的槽钢共同营造了一个与使用者紧密相连可触动感官的创意环境。
3. 塑造新的场所气质——内部表皮瓦解固有的梁柱空间体系。
富有动态感的几何麻绳界面，成为空间的新衣，消解了原有的梁柱空间，让创意办公更富有灵性与活力。麻绳编织作为空间隔断，既限定出各个功能空间的归属感，又因其通透性产生小中见大的空间感受。软装搭配在暖色主调上点缀个性灯饰与休闲家具，力图营造一个舒适并带有情感温度的众创空间。

The office area only has 170 square meters and it's located at QSW Culture Center in Shanghai. Owner wants to create an innovative and attractive working space in a low budget. In this office, small-scale design teams and startups can work together and share resources.
Keywords: share, eco-friendly, open space, material construction

1.Sharing, communicating, creative
The streamlined structure of two rectangles in the open space makes the layout more flexible. The mobile office and meeting office are integrated in the central long table working area, making the space more efficient. The original space below public staircases was too small to be used. According to the space conditions, now this space turns to be a multi-functional area for reading, lecturing and displaying corporate culture.
2.Rope, channel steel, wood
Most materials in this office are natural materials, such as rope, channel steel and wood, which are low cost and eco-friendly. In order to highlight the possibility and beauty of natural rope material, the designer used waving rope as the main design concept. The rough ropes and industrial channel steels establish a creative environment which is closely related to users and simulates their senses.
3.Waving ropes cover beam-column system
Dynamic waving rope interface, likes a new clothes for the space, covers the original beam-column system to make office more energetic. Working as spatial partitions, the rope weaves define the function of each space and produce a space experience that you can see big from small. This office is decorated with warm colors and elaborate lighting and furniture, trying to create a comfortable and friendly space.

274

The Best
Dining
Space
Design Award
最佳餐饮空间设计奖

艾特奖

最佳餐饮空间设计奖
THE BEST DINING SPACE DESIGN AWARD

INTERNATIONAL
SPACE DESIGN
AWARD

获奖者/Winner

杭州设谷装饰设计有限公司（中国杭州）
Hangzhou Shegu Design (Hangzhou, China)

获奖项目/Winning Project

喜来稀肉韩国烤肉餐厅
Seorae Korean Barbecue Restaurant

获奖项目/Winning Project

喜来稀肉韩国烤肉餐厅
Seorae Korean Barbecue Restaurant

设计说明/ Design Illustration

用空间与色彩描绘最纯正的韩式风味

源自韩国首尔瑞草区法式小镇的喜来稀肉韩式烧烤，在地道韩式烤肉根基之上，赋予法式烧烤的精致与匠心。成年猪取300克护心肉，意为"稀肉"，创立至今风靡亚洲各国，客似云来，意为"喜来"。

韩式传统文化所孕育出的"亲民为善，务实求真"使其饮食与建筑两者皆趋向于平民化，简单而富有浪漫。不同文化背景所沉淀出的各类建筑风格中，哥特式风格伟大而神秘，中式风格传统且权贵，韩式建筑则为大众而建，致力于营造平民生活中的务实生活空间，这一理念，成为喜来稀肉韩式烤肉店设计的建筑核心理念。

无论是挑剔的美食家还是平民大众，对饮食的感受，一则是味觉、二则是用餐环境。空间块面分布合理，色调柔和安静，服务贴心快速，成为喜来稀肉韩式烤肉店设计的具体表现方式。

喜来稀肉店内色调平和温暖，沉稳安静的色调，辅以纹理细腻的地砖与墙纸，使其整体氛围更加柔和。铁板镂空、刚毅的水泥地面与墙体，配以恰到好处的柔和灯光。设计师匠心独到，将韩文拆分为细小的关节扣入每一面铁板之中，大小不一的镂空面配以黄晕的光以及金属本身固有的肌理，宛如星空的神秘美感跃然于上。大面积晶莹剔透的玻璃酒瓶在绿色灯光的映衬下显露出绅士般的品质感，沉稳却充满诱惑。木质坐椅及蜡烛的软装点缀，又让空间细腻柔和，整体氛围的营造，如柔软的心包裹在刚毅的外表下，不远不近，点到为止。整体空间格局以块面分割延展，主体空间分割成韩式民宅的剪影效果，并以镂空图形营造多重负空间效果，使其在视觉效果上错落有序，远观通达敞亮，近看精致优雅。空间格局灵活多变，用餐环境更具私密性，用餐者互不干扰，主通道宽敞连通，保证服务的高效性，提升服务品质。

建筑本身既是生活的哲学，而美食则是大众最基本的生活享受。建筑的具象化与美食的味觉感受，有机地结合起来，以建筑本身的形态营造最符合用餐者的就餐环境。大空间块面分割，主体分而不离，色调沉稳柔和，内饰简单精致，从视觉到味觉的和谐共融，才是一个符合生活的优秀设计。

Describe the Most Authentic Korean Flavor with Spaces and Colors

This restaurant originates from the Korean barbecue in Seorae Village, a small French enclave in Banpo-dong, Seocho-gu, Seoul. Its Chinese name "喜来" (Xi Lai) was translated from the pronunciation of "Seorae". Based on the authentic Korean roast meat, Seorae integrates French barbecue's delicacy and ingenuity. It uses 300 grams "shirt meat" from a fully-grown pig, which is so good and so rare, that Seorae has been very popular in Asian countries and attracting a great many of guests since its founding. The traditional culture of Korea gestates the spirit of showing concern for people, seeking truth from facts and being practical, and this kind of spirit makes its dietary and architecture tend to cater more to the public needs, simple yet romantic. Among various kinds of architectural styles under different cultural backgrounds, the Gothic style is grand and mysterious, Chinese style is conventional and aristocratic, while Korean architectures are built for the public, aiming to create practical living spaces for common people, and this is the core design concept of Seorae.

No matter for picky gourmets or common people, flavor and environment are two important things when dining. Seorae has its specific expressions, such as reasonable distribution of spaces, soft colors and quiet tone, intimate and quick service.

In Seorae, the gentle and calm color tone is matched with fine floor tiles and wallpapers, so that the atmosphere is nicer. The soft lighting contributes to a perfect collocation with hollow iron plates, hard cement flooring and walls. The designer inventively split Korean words into several parts and embedded them into iron plates of varying sizes, and the metal texture is illuminated by dim light, making one feel like in a mysterious and beautiful starry sky. Under the green lighting, the translucent glass bottles look tempting. The embellishment of wood seats and candles adds an exquisite feeling in it. The main area is divided to achieve an impression of Korean house, and the hollow patterns create an effect that several negative spaces are visually staggered. Seeing from a distance, the whole space is bright and spacious, and if you look closely, each detail is exquisite and elegant. The flexible layout provides a private dining environment which enables guests to have dinner without being interfered by others. The main passage is spacious and fluent, which could ensure high efficiency and improve the quality of service.

Architecture is a philosophy of life, and fine food is the basic amenity for common people. This restaurant combines architectural form with taste perception to create the most appropriate dining environment for diners. The main large area is divided but not separated away. The color is calm and soft, and the decorations are simple yet refined. A design which harmoniously integrates sight and taste together is an excellent one that meets the needs of life.

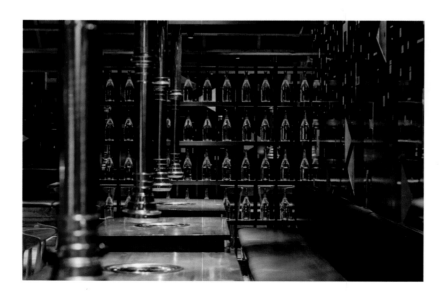

获奖评语

把一种很特别的意念放在一个餐饮空间里，让你走进那个世界去享受美食，是一种很现代感，同时又有传统韵味的一种设计。

Putting a very special idea into a dining space, which lets you walk into that world to enjoy delicious food, is a kind of very modern design but at the same time has traditional charm.

NOMINATION FOR THE BEST DINING SPACE DESIGN

最佳餐饮空间设计奖提名奖

Marina and Alfredo Manca
(意大利/Italy)

获奖项目/Winning Project

Ridola咖啡馆
Ridola Caffè

设计说明/ Design Illustration

"在我眼中Ridola咖啡馆是人们思考聚会的场所，一个文化聚会的场地。它提供一种对所有人来说共同的思路，可以简略表达为我提倡的慢生活理念。咖啡馆内的节奏缓慢，人们不仅有时间去品味食物，也能够品味图画、声音和谈话。这种概念是咖啡馆的基本哲学。"马泰拉这座城市有着复杂的历史和丰富的传统文化。"传统"是设计的关键词。设计开始于对古老的卢卡利亚艺术和工艺的研究，这能够精确定义客户需求。位于马泰拉市中心，Ridola咖啡馆小而惬意，馆内空间因为墙上镶嵌的一面反射镜而显得比原来更大。反射镜上装饰着两块屏风，这种呈现和投影给咖啡馆内带去一种生动的、多样的和当代的氛围。然而，使得项目最有特色的元素，是墙上悬挂的各种传统的图案：大罗盘、雕刻的铁盘、手制工艺品等。地板和桌椅都采用明亮的柚木，使馆内洋溢着一种温馨感恩的氛围。馆内还装饰着大型现代灯具。仅用两种材料——金属和木材，即与原来馆内凝灰岩结构和谐地相互点缀。

"The Caffè Ridola in my view is a meeting place for thinking people, a line of thought common to all that this place offers and which could briefly coincide with the concept, which is dear to me, of living slowly. The pace inside the Ridola is slow and guests have time to reflect on what they savor, not only food but also images, sounds and words. This is the concept that is the basis of the philosophy of the place". Matera, a locality with complex history and rich tradition. "Tradition", not surprisingly, is the key word in the design, which starts from a research on the ancient Lucanian arts and crafts, and can identify the precise needs of the client. Located in the heart of Matera, the "Ridola" is a small and cozy cafè whose perception has been expanded with the help of a reflective glass on one of the short walls that surround the environment. Such mirrored glass has allowed to mask two screens that contribute to give a lively, dynamic and contemporary atmosphere to the place with their immaterial presence and projection images. However, what characterizes the project is the treatment of the surfaces of walls which is entrusted with the task of reinterpreting the message conveyed by tradition; large rosettes, carved iron plat, handicrafts and so on. The atmosphere is warm and welcoming thanks to the use of light oak wood, used for flooring, for covering the counter and the tables; and it is decorated with large contemporary lanterns. Only two materials, metal and wood to bring out the original structure in tuff left visible in the vault and emerging here and there on the walls.

NOMINATION FOR THE BEST DINING SPACE DESIGN

最佳餐饮空间设计奖提名奖

FTC室内设计公司（中国成都）
FTC Interior Design (Chengdu, China)

获奖项目/Winning Project

流浪地图融合菜餐厅
Wandering Map Fusion Restaurant

设计说明/ Design Illustration

当水泥钢板或者镀层钢管盛行在顶部天花板，营造一个粗犷又不受约束的空间氛围的时候，适当改变下设计手法，在顶部运用特别的材质，便可营造非凡的空间氛围。本案整个空间最大的亮点在天花板的构造上，运用钢网层叠穿插，帷幔轻柔之间具有力度美，同时色彩深浅的过渡中形成一幅具有泼墨画的效应，隐约之间形成仗剑天涯的英气，又不失浪漫感。设计师运用巧妙的设计手法使得顶面别有一番趣味，好似别有用心又酣畅淋漓的肆意挥洒，形成的一幅立体的泼墨画。墙体运用两种颜色相互靠近，与顶面的泼墨画交相呼应，保持整个空间的协调统一。运用线条纤细而流畅的木质桌椅，使得空间质感更轻盈，在用餐区域营造了一种舒适自然、轻松愉快的用餐环境。

When cement plate or plating steel pipe prevails on the top of the ceiling and creates an atmosphere of wildness, to change the design method properly, and use special materials on the top place, which could create an extraordinary atmosphere of space. The most character of this project is in the ceiling structure. We use steel net that crosses in layers, and there is beauty of strength in softness, forming an effect of splashing ink painting in the transition of color depth, with vaguely heroic spirit and romance. Designers use the delicate designing method to make the ceiling special, forming a three-dimensional splashing ink painting. The wall uses two close colors to echo with the splashing ink painting on the ceiling, keeping the whole space unified. The wooden tables and chairs with slim and smooth lines make the space quality lighter, and create a natural, comfortable and joyful environment in the dining area.

NOMINATION FOR THE BEST DINING SPACE DESIGN

最佳餐饮空间设计奖提名奖

徐代恒设计事务所（中国南宁）
X.D.H Design (Nanning, China)

获奖项目/Winning Project

沙拉市集
Fresh Salad Fair

设计说明/ Design Illustration

整个空间以优雅大气的冷灰墨色为空间主调，但它带来了一种时尚。在布局上，长方形盒子的小小空间被黑色水曲柳木纹和花砖分布，减少了空间的狭长感觉。天花到墙面"7"字形的花砖铺贴，如一块亮丽的旗帜突显了空间高度，更有霓虹灯缔造的LOGO与之映衬。黑色水曲柳百叶间夹着暖色灯带，增添动感视觉。
木纹装饰吧台，搭配从天花板垂落而下的拉丝钛金管和金属射灯。满排金属厨具装饰，点出年轻、时尚。用餐区的金属吊灯像是音符，给市集带来节奏感。

The key tone of the whole space is elegant cold gray ink color, while at the same time it also brings a kind of fashionable sense. In the layout, black ash wood and tiles are distributed to the little space on rectangle box, which decreases the long and narrow feeling of space. The "7" shape tiles are paved from ceiling to wall surface, like a bright flag that spotlights the space height together with a LOGO made of neon lamp. Black ash tree shutter with warm lighting add a kind of dynamic vision to the space.
Decorative wood bar matches with brushed titanium tube. A metal lamp falling from ceiling. Metal kitchenware decoration displayed on rows shows a sense of youth and fashion. Metal ceiling lamps in the kitchen area are like musical notes, bringing a sense of rhythm to the fair.

NOMINATION FOR THE BEST DINING SPACE DESIGN

最佳餐饮空间设计奖提名奖

HHD假日东方国际设计机构（中国深圳）
Hong Kong Hong Zhongxuan Design Group
(Shenzhen, China)

获奖项目/Winning Project

兰亭京都
Lanting Kyoto Cuisine

设计说明/ Design Illustration

空间设计中运用中国著名书法家王羲之的《兰亭序》作为设计理念，用信息时代的时尚语言重新表达，把书法文字打散，只取其笔划的神与情，书法贯穿于整个餐厅的设计中，如以"天下第一行书"《兰亭序》的笔调作为天花的设计元素。另外，大量运用竹子、原木、原石和水等自然元素营造禅意安宁的氛围。

In space design we use the Orchid Pavilion Preface by famous Chinese calligrapher Wang Xizhi as the design concept, and the fashion language of information age is used to re-express it. We break up the texts and only take the looks and emotions of its strokes. The calligraphy runs through the design of the whole kitchen, for instance we use the style of Orchid Pavilion Preface, which is raised as the first running script (in Chinese calligraphy) of the world, as the design element of ceiling. What's more, plenty of bamboos, logs, rough stones, and water are used to create an atmosphere of Zen and quietness.

298

The
Best
Entertainment
Space Design Award
最佳娱乐空间设计奖

艾特奖

最佳娱乐空间设计奖
THE BEST ENTERTAINMENT
SPACE DESIGN AWARD

INTERNATIONAL
SPACE DESIGN
AWARD

零 壹 城 市

LYCS ARCHITECTURE

获奖者/Winner

零壹城市建筑事务所

（中国杭州）

LYCS Architecture (Hangzhou, China)

获奖项目/Winning Project

V+ Lounge

获奖项目/Winning Project

V+ Lounge

设计说明/ Design Illustration

杭州V+ Lounge是著名品牌V+继北京、成都后国内的第三家分店。项目地处西湖旁最为繁华的核心商业区域，建筑三层以上便可远眺湖景，离苹果亚洲最大的旗舰店不足100米。项目独特的地理位置传递出关于湖水的灵感，设计团队通过对曲线、镜面与光线的运用，将水下、水面、水上的不同感官体验融入空间设计中，创新性地将西湖之景引入室内。
建筑内五层空间融入了湖底漫游的概念，包间与走廊的墙面采用了镜面加流线型的灰色墙体表现深入湖底的水下世界。走廊界面极尽简洁，光藏匿于镜面玻璃后和曲线墙体竖向切片的缝隙中，西湖的远景在这片暗色的涟漪中格外鲜亮。
六层空间引入了无限西湖的概念，空间内沿用了白色的流线型墙体，并与镜面材料无缝连接，纯净而明快。整个空间简约而富有动感，宛如西湖水面一叶扁舟划过后留下的优美曲线。
包间内的设计，为了将西湖的景观价值最大化，每个空间都采取了内外呼应的设计方式。室内使用了大量的落地玻璃，并且部分区域内的顶面和地面加入了光滑的镜面，通过反射能将西湖的美景引入室内，使用户在自然景观的环抱下享受歌唱的乐趣。

The interior space of Hangzhou V+ Lounge, which serves as the famous brand's third branch in China after Beijing and Chengdu, is located at the most flourishing commercial zone nearby West Lake, where you can enjoy the lake view above the third floor. Meanwhile, the site is less than one hundred meters away from the largest flagship store of Apple in Asia. It is this unique location of the project that initiated the inspiration of translating the water of West Lake into a design concept. Curves, mirrors and different lights are adopted to support this concept, thus integrating diverse sensory experiences of underwater, water surface as well as over-water in a whole interior space. It's a creative design which introduces the scenery of West Lake indoor.
The fifth floor is designed with the

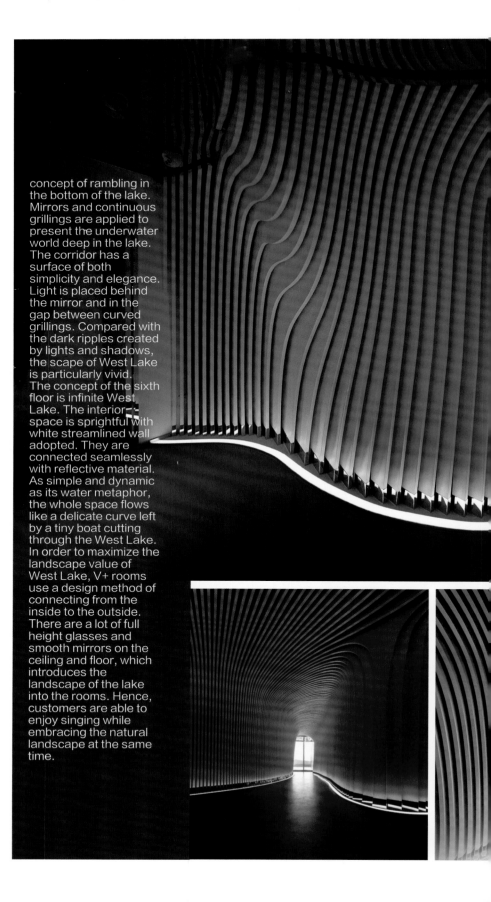

concept of rambling in the bottom of the lake. Mirrors and continuous grillings are applied to present the underwater world deep in the lake. The corridor has a surface of both simplicity and elegance. Light is placed behind the mirror and in the gap between curved grillings. Compared with the dark ripples created by lights and shadows, the scape of West Lake is particularly vivid.
The concept of the sixth floor is infinite West Lake. The interior space is sprightful with white streamlined wall adopted. They are connected seamlessly with reflective material. As simple and dynamic as its water metaphor, the whole space flows like a delicate curve left by a tiny boat cutting through the West Lake. In order to maximize the landscape value of West Lake, V+ rooms use a design method of connecting from the inside to the outside. There are a lot of full height glasses and smooth mirrors on the ceiling and floor, which introduces the landscape of the lake into the rooms. Hence, customers are able to enjoy singing while embracing the natural landscape at the same time.

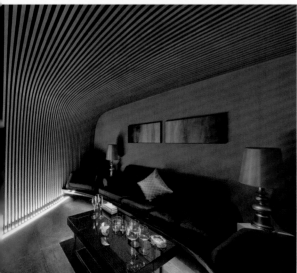

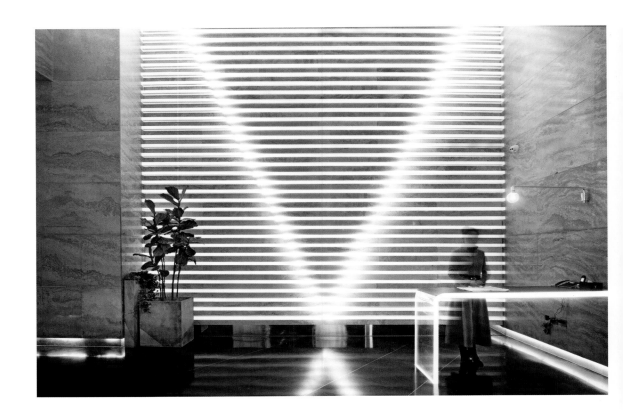

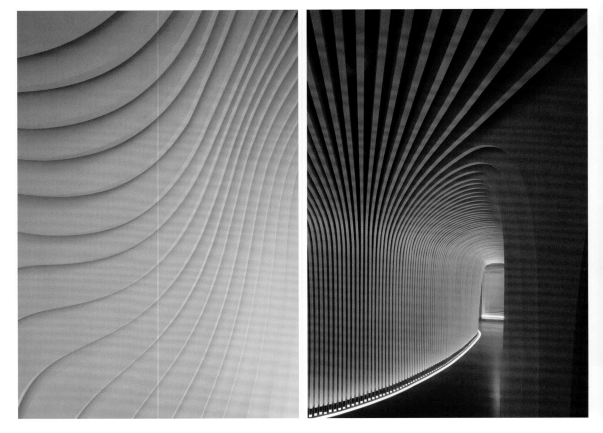

获奖评语

既有活跃的积极氛围，同时也处理得比较简洁，巧妙地把这两者结合起来，它是个比较成功的设计。

Delicately combining both active and positive atmosphere with relatively simple design method, it is a successful design.

NOMINATION FOR THE BEST ENTERTAINMENT SPACE DESIGN

最佳娱乐空间设计奖提名奖

陈武（中国深圳）
Chen Wu (Shenzhen, China)

获奖项目/Winning Project

杰克酒吧
Jack Club

设计说明/ Design Illustration

杰克酒吧位于江苏吴江新城区商业中心市体育场及市政府行政办公中心区域。1000平方米的娱乐空间，耗资2000万元打造，成就了最具江南特色的高档商业娱乐场所之一，设计师在杰克酒吧设计上不只是以风格作为规划主题的首要表现，而是揣摩经营者对于这个空间的热诚与期待，继而成为创意发想的元素与能量，在掌握到的经营者模式下，将苏派建筑风格传统元素，镂刻于立面上，作为与企业精神的研发与解构，完整呼应，衍生客制化的专属魅力。

在消费呈现饱和甚至带有浮躁心理的时代下，稀少的就是极具优势的，中式酒吧的出现恰好弥补了市场这一空缺，并很好地吻合了部分人群追求怀旧的心理。杰克酒吧在一定程度上满足了人们追求品位和内涵的心理。习惯了酒吧光怪陆离，人们对于中式低调独特的神秘氛围抱以好奇，设计师将苏派建筑风格特色运用至杰克设计之中，有的是满满的江南园林Feel，一扇扇镂空雕花屏风、圆形洞门、烛台吊灯、鸟笼式的舞台……为酒吧神秘中增加时空交错的氛围。

设计师将杰克营造出的那种儒雅的文化环境与酒吧娱乐融合，体现出高层次的审美与文化修养，既吻合现代人娱乐需求又能充分体现传统中式的典雅风味。室内软装家具以中

褐色真皮沙发配上传统中国红灯光加之传统元素屏风隔断的设计，很好的避免了古典中式风格所带来的沉闷压抑之感，让酒吧里的传统元素更为协调。

相信只要来过杰克的消费者们定会对杰克酒吧所营造出来的典雅气息赞不绝口。

Jack Club is located in the CBD of the new district in Wujiang, Suzhou, Jiangsu and surrounded by the city stadium and municipal administration center. With an area of over 1000 m², it costs 20 million RMB and becomes one of the most distinctive high-end entertainment arenas which shows features of Jiangnan areas.

Designers not only set the style of Jack Club, but also considered how to operate this nightclub. Therefore, Suzhou Style Architecture could be deeply shown in this case. NEWERA DESIGN totally achieves profession and customization. Nowadays, people live in a fast-paced society. It seems like everything is excessive. In this case, less is more. Chinese style bar is the "less" one. It not only occupies vacancy of market, but also caters to nostalgic mentality and classic taste. Suzhou Style is applied in this case, so people can enjoy the beautiful Jiangnan style gardens through carving screens, round portals, candlestick droplights, "cage-like stage"...

Jack Club links culture and entertainment together, satisfies people who both need aesthetic appreciation and entertainment life. Designers utilize brown leather sofas and traditional screens to break the sense of oppression of old-fashioned Chinese style decoration. Everything looks harmonious in Jack Club.

People who come to Jack Club would appreciate the elegant atmosphere for sure!

NOMINATION FOR THE BEST ENTERTAINMENT SPACE DESIGN

最佳娱乐空间设计奖提名奖

蒋国兴（中国苏州）
Jiang Guoxing (Suzhou, China)

获奖项目/Winning Project

松山行足道
Song Shan Xing Foot Massage

设计说明/ Design Illustration

在中国的历史长河中不乏名人靠足浴养生保健的故事：唐朝一代美女杨贵妃靠经常足浴来养颜美容；宋朝大文豪苏东坡每晚都运用足浴来强身健体；清代名臣曾国藩更是视"读书"、"早起"和"足浴保健"为其人生的三大得意之举；近代京城名医施今墨也是每晚必用花椒水来泡脚养生。可见足浴在中华养生保健历史中占有举足轻重的地位。

本案立足人的需求，倾力打造一个轻松舒适的中式休闲空间。没有过多的色彩修饰，没有大量的造型堆叠，一切还原材质本身的面貌，清新脱俗，不染红尘。光线、气味、声音稍纵即逝，设计师却通过对各种触觉、视觉、嗅觉的切割、糅合，带你进入一个新的国度，极具魅力。

大厅墙面运用竹编并黑色压条有序地分割；吧台背景采用斧刀石，自然的肌理，天然的质感，一切回归自然；吧台运用鱼鳞状格栅，造型在灯光的照射下，显得格外的耀眼。还运用红色的瓦砖点缀，让色彩变得鲜亮明快。整个公共区域地面采用黑金沙、山西黑、中国黑三种颜色石材拼花，体现一种沉稳、低调。门采用的是木拼条通顶的设计手法，延伸空间的尺度。过道墙面采用水泥砖用原木色压条有序地分割，其中掺杂着竹林景观，灯光透过

竹林，整体空间活泼而有序。

包间顶部的竹条编织，别出心裁，光束透过间隙落下来，层层叠叠，让人着迷。简洁的线条，铿锵有力，使得空间内有了力量。棋牌室精致的灯笼吊灯，点亮了房间，映着江景，分外别致。墙面用水墨风格的装饰画点缀，写意不写实，给人以超脱万物，置身于仙山灵水的感觉。最简单的魅力所在，更高深的东西，只能靠自己去摸索，去体会……

茶室设计取自中国古文化，横平竖直、方方正正、刚正不阿，并设置竹林景观，竹枝杆挺拔、修长、亭亭玉立、袅娜多姿，四时青翠、凌霜傲雨，竹子具有"宁折不弯"的豪气和"中通外直"的度量，它性质朴而淳厚，品清奇而典雅，形文静而怡然，所谓"未出土时已有节，待到凌云更虚心。"纤细优雅，静谧，身处这里，不忍大声喧哗，怕惊扰这美好。

红酒吧顶面采用镜面马赛克，配合灯光的运用，低调奢华的氛围，让人感觉到更多的是神秘色彩。洗手间延续了过道的设计手法，体现出整体、统一性。卫生间墙面使用黑色线条分割，配以竹节砖，竹节有着节节高升、步步高升的寓意。干练之余，不至拘谨。

竹，秀逸有神韵，纤细柔美，长青不败，高风亮节，高尚不俗，生机盎然，蓬勃向上……它有许许多多的优点，然而，我最欣赏它的坚贞不屈。它那"孤生崖谷间，有此凌云气"的美好品质。它偃而犹起、柔中有刚的高尚品德时时刻刻激励着我！

Since Tang dynasty, foot bathing has been popular as a way to keep healthy and live longer. Yang Guifei, one of the Four

Beauties of ancient China, maintained her beauty by foot bathing; Su Tungpo, a great Chinese writer and poet in Song dynasty, built his body by foot bathing; Zeng Guofan, a famous politician in Qing dynasty, had three habits, getting up early, reading, and food bathing, which have been kept in his whole life; Shi Jinmo, a renowned expert and practitioner of Chinese traditional medicine, used pepper water for foot bathing every day. From these historical stories, we can know that foot bathing is very important.

In this case, the designer hoped to create a Chinese style leisure space on the base of human needs. Without using too many colors or shapes, the designer recombined different feelings, smells and visions to make an amazing space.

There are two attractive sights in the hall, one is the wall paved by black bamboo weaving orderly; another one is the counter which applies fish scale grids, under the irradiation of lamplight, it looks very shiny, and the red bricks makes the color more brightly.

In the public area, the parquet floor is paved by marbles in galaxy black, Shanxi black and China black, steady and peaceful. The door is made of full wood puzzle to extend the space, and the wall is built by cement bricks with wood bead to separate in order, along the passage you can see bamboos, through which the light is reflected on the wall. The whole space is vividly and orderly.

The bamboo weaves on the top of the room are unique, the light shines through the gaps of the weaves, charming and attractive, and these simple lines also power the space. Against the river, the delicate lantern droplights illuminate the chess room. Walls are decorated with traditional Chinese ink paintings, which enable the viewers to feel like in heaven. You need to explore and feel the charm and profound meaning by yourself.

The design idea of tea room comes from China's ancient culture, and it's decorated by rectangular and square shapes. In China, bamboo is not just a plant, but it's often used to represent a style of life. It's thin but long and straight, and green all year, its

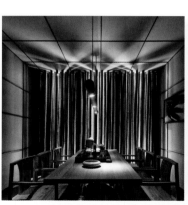

character "rather break than bend" always describes the personality of someone. The bamboos create an elegant, pleasant and peaceful atmosphere, which enables you to enjoy a great time.

In the bar, the ceiling of mosaic mirror reflects the light to many directions, which make the atmosphere luxurious and mysterious. In washroom, the decoration extends the design of the aisle to keep the overall unity of the whole space. The wall is split by black lines and paved by bamboo tiles, implying the business will be more and more successful.

Bamboo is especially appreciated by people because it's slender, feminine, evergreen, ethical, noble, vibrant, vigorous, etc. However, the designer enjoys its unyielding character. Its good quality that "even growing lonely in the valley, it still has a strong ambition to reach the clouds" encourages him and its powerful but gentle feature always inspires him in his life!

NOMINATION FOR THE BEST ENTERTAINMENT SPACE DESIGN

最佳娱乐空间设计奖提名奖

徐麟（中国沈阳）
Xu Lin (Shenyang, China)

获奖项目/Winning Project

银河国际俱乐部
Yin He KTV

设计说明/ Design Illustration

阑珊灯火映照的苍茫夜空，浮云如被点燃的青烟，欧洲街景风格环境很容易让人沉浸在一种对历史的怀感中。复古的大理石、精雕细琢的高大壁炉是最能够吸引人眼球的焦点，还有炉台上极富艺术感染力的雕塑作品、大幅油画，整个壁炉更像是一件精心设计的艺术佳作。柔软的高背沙发也为大堂增添了温暖的气息。明亮的街灯与发光的玻璃遥相辉映。匠心的细节、艳丽的色彩氛围，随时可以沾上你记忆的颜色！以颜色为切入点，彩绘玻璃、复古红砖都直白地表现了欧式复古风格的独特魅力。用色彩表现空间的张力，再加以灯光音频来达到独特的视觉效果，不同的色彩符号，糅合新颖的造型，以独特的审美角度打造独树的KTV风格。主题包房风格迥异，其中有米奇、灰姑娘、法拉利、Dior等主题包房，在环境气氛的延续下更加突出风格的多变性，匠心的细节、艳丽的色彩氛围，无时无刻不在吟唱着其独特的韵味。

Lights illuminate the boundless sky where you can see the clouds look like smoke, an European style street could easily let people to be immersed in memory of history. The tall fireplace made by exquisitely carved marbles gives a retro flavor and becomes the most attractive element, and the sculptures and large oil paintings on the base create a strong artistic appeal, making the fireplace more like a well-designed artwork. The high-backed sofas nearby also make people feel warm. The bright lamps on the street and glittering glasses add radiance and beauty to each other. In a colorful and exquisite environment, it's easy for you to find a color which can take you back to the past. The stained glasses and antique red bricks express the distinct charm of European classical style in a straight way. The designer blended colors and lighting together to accomplish a unique visual impact, and managed to create a special KTV style by using different colors and new structures. Each room has its own features based on varied themes, such as Mickey Mouse, Cinderella, Ferrari and Dior, besides, such variety is more prominent under the detailed and gorgeous atmosphere, and each place is singing a specific song all the time.

NOMINATION FOR THE BEST ENTERTAINMENT SPACE DESIGN

最佳娱乐空间设计奖提名奖

成都旷石室内设计有限公司 (中国成都)
VASTONE Design (Chengdu, China)

获奖项目/Winning Project

成都Wizar原创鸡尾酒吧
Wizar Bar

设计说明/ Design Illustration

《了不起的盖茨比》中充斥着调酒与东方情调的场景，人们调酒如东方法师调制秘药，饮酒如服秘药，坠入迷离。那是20世纪20年代的爵士时代，经典鸡尾酒诞生的禁酒令时期，鸡尾酒与东方美学前所未有的融合，谦和又有历史感，却总能激起你的好奇。这是Wizar Bar设计的基调氛围。

酒吧神秘的入口引导客人进门后瞬间浸入雅痞的热闹氛围，但异形空间、三面临街落地窗、商场照明光污染都成为设计speak easy酒吧的问题。因此，我们更改酒吧入口，以吧台核心区域，依消费者习惯设置私密散座区，在临窗区域设交社交散座区；设可调节角度的格栅与屏风来进行有节奏的遮挡。保证私密与安全的同时，由外向内看是髯松疏影，光影斑驳。一灯一几，或定制，或原创，呈现东西方恰当融合的空间。

主入口以火烧板为基础铺贴，拉丝金属收边，传递神秘而精致的感受。主入口门使用原创纹理进行雕刻，扶手处雕刻酒吧标识符号。进门则窥见吧台一角与酒柜背景形成的景致，其中水滴灯亦取自该画家作品，客户要求吊顶不做处理，水滴灯在此分割空间来解决吊顶过高所带来的就坐不适及不安，兼顾美感。

私密散座区原本小而闷，却适合进行私密交谈；

茶镜的加入满足私密需求又不压抑。布艺沙发与柔和壁灯，放松消费者感官，烘托亲密交谈氛围。Wizar Bar的最终呈现是美的，美不是漂亮、好看这样肤浅，而是独特的东方魅力给人们带来便利与欣然。我们很荣幸地给这家鸡尾酒吧取了名字，Wizar Bar—东方秘法师的酒吧。

The Great Gatsby is full of scenes of mixed drink and oriental exoticism. People mixing drinks is like oriental masters mixing medicine. Drinking is like taking medicine, and people fall blurred. It was Jazz Age in 1920s. Classic cocktail was invented in Prohibition Era. The combination of cocktail and oriental aesthetics is unprecedented, without overindulgence, gentle and historical, which could always arouse your curiosity. This is the basic tone and atmosphere of Wizar Bar.

The mysterious entrance of the bar leads guests to immediately sink into the yuppie and exciting atmosphere. However the alien space, the three French windows facing the street and the light pollution from the mall lighting are all problems of designing a "speak easy" bar. Therefore we redesigned the bar entrance. The bar counter is in the core area, and we set private seats according to the habit of consumers and social seats in window area. We set angle-adjustable grates and screens to shade rhythmically, guaranteeing privacy and safety. Seeing from the outside toward inside is mottled shadow. The lamps are custom or originally designed, showing the appropriate space integration of east and west.

The flamed granite is the basic paving of main entrance, with brushed metal on the edge, conveying a kind of mysterious and delicate feeling. The main entrance door uses original lines to sculpt, and there is a bar sign sculpture in the handrail. Scene formed by a corner of bar counter and wine counter background could be seen when people enter. The water drop lamp is an artwork by a painter. The client asks us not to deal with the suspended ceiling. The water drop lamp in this partition space is to soften the discomfort of sitting and also add a sense of beauty.

Private seats are originally small and quiet, which are suitable for private conversation. The design of the tea mirror satisfies private demand without being depressing. Fabric sofa and soft wall lamp relax customer's senses, creating an atmosphere of intimacy. The final presenting of Wizar Bar is beauty. Beauty is not as shallow as beautiful or good–looking, but special oriental charm that brings people convenience and joy. We are honored to name this bar Wizar Bar––the oriental mysterious master's bar.

322

The
Best
Hotel
Design Award
最佳酒店设计奖

艾特奖

最佳酒店设计奖
THE BEST HOTEL DESIGN AWARD

INTERNATIONAL
SPACE DESIGN
AWARD

获奖者/ Winner

上海鲲誉建筑装饰设计
有限公司（中国上海）
KNYU Construction (Shanghai, China)
获奖项目/Winning Project
澍德堂/Su House

获奖项目/Winning Project

澍德堂
Su House

设计说明/ Design Illustration

"问余何意栖碧山，笑而不答心自闲。桃花流水窅然去，别有天地非人间。"位于安徽的澍德堂有1300年的历史，古村内有一12000平方米的荷花塘，澍德堂虽大，但为了保证品质感，能容纳量只做了32人。外部修旧如旧，如果不仔细看，好像就没修缮过，把时间的印记和岁月的沧桑都保留住。说到这里要感谢当地的老工匠（以老二哥为代表），他们对徽派建筑的理解和热爱才成就了这如画卷般的建筑外观。

内部进行了一场大型"内科手术"，脱胎换骨式地改造，原有的木结构都更换成了钢结构和混凝土；对老宅阴暗潮湿、不隔音等缺点进行了很好的规避；对所有现代化的设备进行了巧妙的隐藏，以符合现代人的生活起居；风格上尽量简单。在这样的环境里，装饰是多余的，时间是凝固的，在质朴简单中细细感受时间的印记。

酒店的入口是最花心思的，也进行了多次调整，目的就是怎样去酒店化，让你走进来的那一刻就融化到画卷里。最得意的地方是餐厅，也是澍德堂的核心区域，也是客人的交流区域，更是和时间、自然交流的地方。我所做的就是定格住时间，把原汁原味的东西加以凝固让老建筑注入新的灵魂。

"You ask me why in the green mountains I reside. I smile and would not answer, my heart is at peace. On quiet flowing streams, peach-blossoms gently ride, in this unworldly place where all things are at ease. " (This is the poem "Question and Answer in the Mountains" by Chinese poet Li Bai.) Su House, located in Anhui, has a history of 1300 years. In the old village, there is a 12000 square meters lotus pond. To ensure high quality service, it only provides accommodation for 32 guests though it's very large. The tracks of time are shown on the exteriors, and if you don't look closely, it seems that they haven't been renovated. Thanks to the local craftsmen, it's their comprehension and love of Hui style architecture that accomplish the picturesque architectural appearance. The interiors have been modified

completely. All the original wood structures were replaced by steels and concretes, which avoided the disadvantages like darkness, damp and bad sound insulation. In order to accord with the living style of modern people, the design firm cleverly hid the modern equipments. Ornaments are unnecessary in such an environment, time stands still and you can feel it in this plain and simple place. The entrance of the hotel required the attention most, and the goal was to make you feel that you are entering into a picture, so the design firm has made lots of adjustments. The best part of the project is the dining room, which is also the core area for communication among guests, time and nature. The designer froze the time by using original items and added new spirit to the old architecture.

获奖评语

非常朴实，但是又不缺它的细腻，特别是它那种沧桑的感觉，体现了一种人文的精神，是个非常成功的设计。

This is very plain, however does not lack its delicacy, especially a feeling of vicissitudes in it, reflecting a kind of humanity spirit, which is a very successful design.

NOMINATION FOR THE BEST HOTEL DESIGN
最佳酒店设计奖提名奖

FCD浮尘设计工作室 (中国苏州)
FCD (Suzhou, China)

获奖项目/Winning Project

浮点·禅隐客栈
Floating Point · Zen Hidden Inn

设计说明/ Design Illustration

"浮点·禅隐客栈"是由一栋老宅改建而成，改造前的"浮点·禅隐客栈"是锦溪古镇南大街上2幢毫不起眼的破房子，老屋门前荒草丛生。曾经的白墙也在雨水的冲刷中变得斑驳，破败感中带有年代的气息。在拆建的过程中，设计师在保留老房子灵魂和神韵的基础上进行内部的设计与改造，希望走进来的每个人都可以感受到人文与设计相结合的意境，以及当地浓浓的风情。

建筑主材选用：青砖及瓦片、H型钢、竹子、白水泥、老木头、通电雾化玻璃等。还有些是就地取材、保证资源的循环再利用。

在建筑外观上屋顶选用青瓦，利用拼接工艺将瓦片延伸到墙面，让我们的建筑更简约，但又保留了江南水乡的建筑特点。在此项目中室内外运用了大量的竹枝、竹桠作为装饰，从而将禅境中乡野、荒蛮的意境完全体现出来。设计中选用竹子的原因是，竹子造价低，又很容易让人感受到禅的韵味、意境。

内部空间布局：客栈分为3层，9间客房，每间客房都有独有的特点。通过精心布置，有美有意境。开放式的空间布局，现代与复古的交融碰撞，白色墙面与浅色地板的辉映，精挑细选的简约设计家具，还有那唯美的纱幔垂于各处，每一处线条和灯光都十分考究。客房和公共区域随处可见的席地座榻，可看茶，可冥想，独守一份禅静。

The "Floating Point · Zen Hidden Inn" was renovated from two old shabby houses located in the south avenue of Jinxi Ancient Town. After years of rain-washing, previous walls become mottled and exude an antique flavor. During the construction, the design team modified the interior space on the basis of the soul and spirit of the old houses, hoping that people can not only feel the artistic conception of humanity and design, but also experience the strong local style as soon as they walk in.

Various materials such as bricks, tiles, H type steels, bamboos, white cements, old woods and electric switchable glasses were applied in the project. The team also obtained some materials from local area so as to ensure the recycling use of resources. The team spliced different grey tiles from the roof to the walls, and such fa ade not only makes the architecture simpler, but maintains the characteristic of Jiangnan style. The project used lots of bamboo branches and forks for decoration, which present the rural wild prospect expressed in the realm of Zen. The design team selected bamboo because of its low cost and ability to make people feel the flavor of Zen style easily. This three story inn has 9 guest rooms. After carefully arrangement, each one is beautiful with its own features. The inn has an open layout that integrates modern and classical elements. The white walls and light color floors add radiance and beauty to each other, the simple style furniture is carefully chosen by the designer, and the curtains hanging in different spaces are graceful, besides, lines and lighting in each space are very well-designed. There are lots of seats in the guest rooms and public areas where you can drink tea, think or enjoy the quietness.

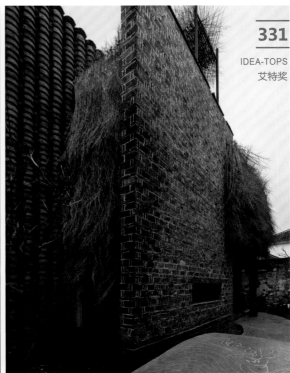

331

IDEA-TOPS
艾特奖

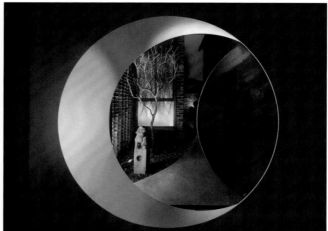

NOMINATION FOR THE BEST HOTEL DESIGN

最佳酒店设计奖提名奖

nA Nakayama Architects
(日本/Japan)

获奖项目/Winning Project

坐忘林
ZABORIN

设计说明/ Design Illustration

"坐忘林"的名字灵感源自于"坐禅"，表达了在大自然的树林中安然静坐，忘却自身的和谐意境。

坐忘林依偎在花园（Hanazono）的一处静谧的私家森林中，这里是新雪谷地区的一部分，四周环绕着北海道未经触碰而广袤的美景，让人感觉远离世俗，但又方便前往附近的滑雪场。坐忘林是一家仅有15间客房的私人日式旅馆，宁静而私密。此处的雪景非常独特，旅馆随着四季的变化与不同的景色融为一体。每间客房都将宁静祥和的日本传统文化体验与所有舒适的现代设施完美结合，提供私人的室内及露天浴池，其温泉水直接来自于坐忘林温泉。该温泉水也被用作能源的可持续性来源。此外，每间客房内都可以欣赏到不同的美景。

坐忘林位于其自然美景的环抱中，将日本传统文化、大自然、菜肴和现代风格的精髓融合在一起，营造出简单而宁静的世外桃源。这是一处体验所有感觉的地方：从风声和水声，到树叶、大地和白雪的味道，质朴美丽而粗旷的景色，感受自然与设计，再到品尝地道的菜肴。这个项目完全可以运用大量的日式材料来设计，但设计师认为坐忘林需要拥有更多的日本精神或灵魂，而这种精神应该通过佛教的"坐忘"来表达。如果采用传统的京都设计，将无法与

法与当地的气候相融，空间就会显得肤浅、不自然。

在设计师看来，"坐忘林"内有"某种东西"，这种"东西"已经被日本人民所遗忘，但它无法用语言来表达。他想要用空间和忘却的意境来营造日式之美，让空间变得富有格调，让人们在幽深、静谧和高雅的世界中感到真正的"活着"。

Inspired by the Zen 'zazen', the name Zaborin merges the traditional Japanese characters, 'zabo' meaning "to sit, to forget" and the character 'rin' representing a small forest or wood.
Zaborin is nestled in the quiet of a private forest in Hanazono, and surrounded by a pristine expanse of Hokkaido landscape, feels remote despite its easy access to the resorts nearby. Zaborin blends a tranquil, traditional Japanese experience with all the modern comforts and amenities.

Zaborin is an intimate ryokan of only 15 villas that reveres tranquility and privacy. The snowy landscape here is particular different, the building coexists with avariety of scenery throughout the year. Each villa has its own private indoor and outdoor bath with spring water fed directly from the Zaborin onsen. The hot spring is also used as a sustainable source of energy. The view is totally different from each room and you are able to enjoy its own beauty.

Amidst the encompassing beauty of its natural surroundings, Zaborin combines all the best of Japanese tradition, nature, cuisine and contemporary style, creating a refuge of simplicity and serenity. A place amongst the trees to sit and to forget, to experience the sound of trees and water, the smell of leaves, earth and snow, the simple but beautiful view, the nature, design and local food.

The designer totally understands it's able to make Kyoto space by using the Japanese material abundantly. But he thinks ZABORIN needs to bring more Japanese spirit/soul, and has to express the meaning "SPIRIT" from Buddhist terminology "ZABO".

ZABORIN cannot be traditional Kyoto design. If do this design in Hokkaido, it will not match the climate, and space becomes superficial and full of artificiality.

He really thinks there is "SOMETHING" in the "ZABORIN", something that Japan and Japanese forgot already. And this "SOMETHING" is beyond description. He wanted to create the Japanese beauty with space and free of all thoughts. Furthermore, he intended to make space with high level and character. He also wants people to feel "alive" in here. Also feel the profound and quiet elegance world.

NOMINATION FOR THE BEST HOTEL DESIGN

最佳酒店设计奖提名奖

刘万彬（中国成都）
Liu Wanbin (Chengdu, China)

获奖项目/Winning Project

空·堂—青城山景泰阁 精品客栈

Empty·Hall—Green City Mountain Jingtaige Boutique Inn

设计说明/ Design Illustration

项目位于青城后山景区内一处荒置的院落建筑，经久闲置，破旧凋敝，经历2008年汶川地震后，多处出现裂缝和材料脱落。内部颓丧的院落也十分荒凉。在满足客栈运营和整体形象需求的前提下，尽可能保存和延续院落建筑的格局和建筑要素，设计侧重于整理和协调，将人与景观的关系作为考量核心，自觉削弱装饰的力量，赋予客栈院落空间更多的接纳和包容感。

"空堂"是对客栈实际效果的准确定义，也是整个设计思想的核心。接近于"背景营造"的设计态度，使多样化的建筑、院落、室内有了共存可能，既尊重建筑本身既有的元素，也有效控制了建设成本。

传统街区风貌中，充满仪式感的门楼嫁接清新放松的内庭，关照了旅客的内心需求，也尊重了投资人严苛的预算控制。客栈内外形式上的反差和矛盾关系，正是功能目的性高度一致的体现。

空山新雨后，天气晚来秋，青城山清爽幽然的天然特质，炎暑之中，让人神往！空·堂便是一处清雅的身心安放之地。

The project is an unused courtyard building located at the back mountain scenic region in green city. It has long been unused and is quite old and poor. After going through Wenchuan earthquake in 2008, there are many cracks and materials shedding.
The interior of the courtyard is also very desolate. On the basis of satisfying requirements of operating and the overall image, the design preserves and continues the architectural layout and elements as much as possible. The design focuses on organizing and coordinating, and the core considerations are the relationship of human and landscape, weakening the decoration consciously and giving the inn courtyard space more sense of acceptance and tolerance. "Empty Hall" is the precise definition of the actual effect of inn, and is also the core concept of design. Close to the design attitude of "background building", the diverse architecture, the courtyard and the interior have the possibility of coexistence, which both respect the elements of the building itself, but also effectively control the construction cost.
In traditional style block, gate house full of sense of ritual connects with fresh and relaxing chamber, which cares for the inner needs of visitors at the same time respects the strict budget control of investors. The contrasts and contradictions of the inner and outer inn are exactly the high embodiment of the functions and its goals.
After raining in the empty valley, it becomes cold as it is in fall. The natural and pleasant features of mountains in Green City are very attractive to people in hot summer. Empty · Hall is an elegant place to relax for both body and mind.

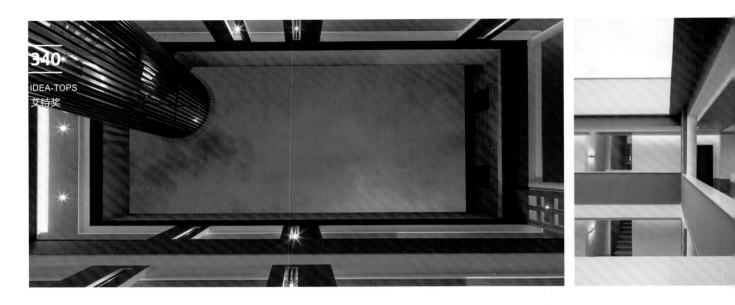

NOMINATION FOR THE BEST HOTEL DESIGN

最佳酒店设计奖提名奖

上海鲲誉建筑装饰设计
有限公司（中国上海）
KNYU Construction (Shanghai, China)

获奖项目/Winning Project

悦庭楠舍酒店
Yue Ting Nan She Hotel

设计说明/ Design Illustration

楠舍前身是一个停用的疗养院，也是一个"中国欧式"建筑。我们认为在楠溪江这个美丽的中国乡村，有更适合这里的建筑形式，因此，利用当地的材料对建筑外观做了比较大的改变。拆除了南面围墙，做了一条美丽的稻田栈道。原来的房间两间合并为一间，房间更大，更舒适。宽大的落地窗保证了充足的采光，也使视野更开阔，躺在床上也能看见远处的风景。隔离出每间客房的阳台，使得客人能尽情享受而不会被隔壁打扰。

之前传统的绿篱花园改为草地，一是视野更开阔，坐在房间里也能看见远处的风景，二是希望来酒店的客人能坐在草地上晒太阳，真切感受自然。宽阔平坦的草地也便于举办很多户外活动。

楠舍很多地方都是老物新用，大木床由老木工师傅选用老房子拆下来的梁柱，用榫卯结构的方法搭建。搭配棉麻的白纱，这种气质和楠舍再符合不过了。在这个项目里，我们看到了老木头的无限潜能，它比新木头性能更稳定，颜色更自然，基本不用调色的油漆，只是轻轻上一遍木蜡油就好，基本没有装修污染。

Nan She Hotel is originally a nursing home that no longer being used, and also a "Chinese European style" architecture. We think in such a beautiful Chinese countryside Nanxi River, there shall be an architectural form that suits here more. Therefore local materials are used to change the appearance of the architecture. South wall was torn down and we designed a beautiful rice field road. The original two-rooms house was changed into a one-room house, which is larger and more comfortable. Wide French window guarantees abundant sunshine as well as wider view, where people can see scenery in the distance lying in bed. The balconies of each guestroom are separated so that guest can fully enjoy the view without being disturbed by neighbors.

The original traditional hedge garden has been modified into lawn, not only providing a wider view, where guest could see scenery in the distance just sitting in the room, but also in the hope that guest could sit on the lawn and enjoy the sunshine, to feel nature vividly. The wide and plain lawn is convenient to hold many outdoor activities. There are many old things being reused in the design of Nan She Hotel. The big wooden bed is made by the old beams and columns chosen by old wood masters, and they build it using the tenon and mortise structure. Matching with white cotton yarn, the temperament could not suit Nan She Hotel more. In this project, we see the unlimited potential of old wood, which performance is more stable than new wood, with more natural color. Basically no need for color paint, just lightly paint wood wax oil, which barely cause pollution.

346

The
Best
Club
Design Award
最佳会所设计奖

艾特奖

最佳会所设计奖
THE BEST CLUB DESIGN
AWARD

INTERNATIONAL
SPACE DESIGN
AWARD

获奖者/Winner

于强室内设计师事务所（中国深圳）
YuQiang & Partners Interior Design (Shenzhen, China)

获奖项目/Winning Project

水湾1979【云端】会所/Shuiwan 1979

獲奖项目/Winning Project

水湾1979【云端】会所
Shuiwan 1979

设计说明/ Design Illustration

"我们都是时间旅行者，为了寻找生命中的光，终其一生，行走在漫长的旅途上。"多少人向往"一次说走就走的旅行"，却不是每个人都有足够的勇气去实现梦想。

立于云端，让梦想如阳光照进现实，如同人们引起共鸣的水湾1979的理念。"You are what you live。"——自我，是生活的反照。

作为水湾1979的战略级合作伙伴，于强室内设计师事务所在接到水湾1979二十四层会所的设计需求后，确定了基于空间本身得天独厚的景观资源，以"云端的旅行"为出发点，打造一个真正的"云端会所"的设计目标。

当二十四层的电梯门开启，清爽的视觉美感、独特的人文气质以及身心的极致舒适，是这个中空层高达8.2米的挑高空间给人的第一印象。一组白色复古旅行箱组合成了接待台，具有艺术气息的家具装点空间的同时给人极致的享受，年轻人钟爱的波普艺术让空间灵动起来……"渐变"既是一种材质上的表现手法，同时也是设计内涵上的诚意表达，"在丰富人文背景的历史土壤里，孕育出新的时尚艺术，正是水湾的发展所带来的变化"。

从会所俯瞰，如同漫步云端，极目远眺之处海天一色，滨海城市的旖旎风景给人无限的视觉体验。当阳光透过整面的玻璃窗随着时间的变化折射出不同光影，你能感受到一旁高大的绿色植物在光合作用下生机盎然，与友人对坐喝一杯咖啡，这一场云端的旅行，心情明亮澄澈。

"We are time travelers who spend our whole life searching for light." Lots of people want to go on a trip at any time, but not everyone has enough courage to fulfill the dream.

The concept of Shuiwan 1979 is to make dream come true. "You are what you live."

As a strategic partner of Shuiwan 1979, YuQiang & Partners Interior Design started from the idea of "travelling through the clouds" based on the advantaged landscape, aiming to create a real "club over the clouds" on the 24th floor.

As soon as the elevator door opens, this 8.2 meters hollow space allows people to experience its fresh look, special humanistic environment and extreme comfort. A set of white vintage suitcases forms the reception desk, the artistic furniture not only embellishes the space, but gives people aesthetic pleasure. The pop art which is popular with young people enlivens the whole place. "Gradual change" is a method of material use, and it's also an expression of connotation, "Shuiwan gestates new fashion art in the historical soil of rich cultural background, this is the change that it brings."

When overlooking from the club, it's like walking in the clouds. The sea and the sky merged into one, and the beautiful sceneries of the coastal city provide infinite visual experience. The sunshine reflects different shadows through glass as time coffee with your friend here, and a travel through the clouds could brighten your mood.

获奖评语

它有很多艺术的表现，还有很多家具安排得非常有序，有点工业风，有艺术的成分在里面，非常成功。

It has much artistic expression, and a lot of furniture is arranged very orderly, with a little kind of industrial style and art composition in it, which is very successful.

NOMINATION FOR THE BEST CLUB DESIGN

最佳会所设计奖提名奖

陈子俊 (中国香港)
Joe Chan (Hong Kong, China)

获奖项目/Winning Project

三亚半岛云邸会所
Sanya Peninsula Club house

设计说明/ Design Illustration

瀧澮（会所名称），位于海南省三亚的半山半岛，此坐拥三山两湾的世界珍罕宝地上，是为中信半岛云邸小业主而设的住客会所，整个会所包含大堂、咖啡厅、中餐厅、包间、品茶室、图书馆、棋牌室、健身房和泳池等空间，总面积超过5000平方米。

甲方要求会所的风格是新中式和具有中国文化底蕴。

本项目在空间规划上做出了非常有效的配置，不仅使很大部分的废置空间（大堂）都得到有效的利用，而且对原来规划的合理调整使空间（户外泳池和咖啡长廊）使用更加合理。此外，用的火山岩石、蒙古黑大理石、烟熏色橡木和爵士白大理石等低成本材料做出的新中式风格，很大程度上体现出设计师"粗料精做"的功力。烟熏色橡木三交六椀菱花格栅连贯多个并排的功能空间，使整体感觉统一和谐。

155米长廊连通图书馆、茶室、讲经堂、中餐厅、健身房等空间，古代的竹雕、瓷器、茶具等古玩突显中国文化的底蕴。"走过红尘的纷扰，弹落灵魂沾染的尘埃，携一抹淡淡的情怀"。

The clubhouse Longhui, located in Peninsula, Sanya in Hainan Province, commands a rare panorama of three mountain ranges and two bays nearby. Exclusively for the residences of CITIC Peninsula Estate, a property developed by CITIC Real Estate in the peninsula. The clubhouse boasts an array of facilities including a lobby, a cafeteria, a Chinese restaurant, private rooms, a tea room, a library, a chess room, a gymnasium and a swimming pool, which has a total area of 5,000 m2.

Client requested the clubhouse to be decorated in a Neo-Chinese style which would feature the essence of Chinese culture. This project is an example of an effective and wise allocation of space, not only in the way that most of the disposed space (which turns out the lobby) is given a new role, but also in the sensible amendment to the original space planning (outdoor swimming pool and coffee lounge). Besides, the use of low cost materials such as volcanic rocks, Mongolian black marbles, smoky oak and jazz white marble for the Neo-Chinese style demonstrates the designers' capability in delivering the finest works even with the roughest materials. The smoky oak grille with a Sanhualiuwan rhombus pattern connects the individual functional rooms with a harmonious coherency in a whole.

The 155 meters long passage connects the library, tea room, sermon hall, Chinese restaurant, gym, etc., along which some decorative items such as bamboo artefacts, porcelains and tea wares showcase the treasure of Chinese culture. "After going through different experiences in the world, we sweep out all the ashes in our soul and maintain a peaceful mind."

NOMINATION FOR THE BEST CLUB DESIGN

最佳会所设计奖提名奖

凌子达（中国上海）
Kris Lin (Shanghai, China)

....

获奖项目/Winning Project

时代南湾会所
Times Ocean Club House

设计说明/ Design Illustration

生活的意义源于对生命中本质的追求，褪去华丽的外衣，摒弃一切的浮夸后，最终还是回归自我心中的价值，回归生活的真谛，而"艺术"是人类对生命不停的探索，反复的辩论与反思，是对生活的实践与印证，最终达成思想，完成艺术作品。"生活的艺术，艺术的生活"是本案设计的核心价值，打造一个具有艺术气质的会所。

不像绘画，是平面式（2D）的呈现，而雕塑更有立体构成，雕塑家是由三向度来思考与完成作品，本案试图以雕塑的手法来完成空间的设计，突破2D（平面）的思考模式，全部3D建模来设计，由"点"拉成一条"线"，再由"线"组成一个面，由许多不同角度的"面"组构成立体的空间。每一根线条，每一个面，就像雕刻家一样一刀一刀精准地切割出来，最后完成的室内空间本身就是一个艺术品。

透过几何线条和灯光的线条，加上大理石马赛克的纹路和原木色彩相结合，彰显大堂空间的时尚休闲感。LED灯激发出的柔和光色，营造出咖啡区独特的优雅休闲氛围。楼梯本身就是一件雕塑品，并透过不规则的线条，使楼梯更富层次感。游泳池以海浪为主题，表现出天花吊顶的层次感。健身区包含了健身房、舞蹈教室、台球室、乒乓球室等

功能区。篮球场内将墙面与天花相结合，形成一种优美的曲线关系，同时结合LED灯的方式在整个空间产生一种波浪形的层次感。

The meaning of life originates from the pursuit of the essence in life. Taking off the gorgeous cover and abandoning all the pomposities, it finally goes back to the value in people's heart and the truth of life. But art is unceasing exploration for human life, the repeated debating and rethinking, the practice and verification of life, which finally concludes as an idea and completes the art works. "The art of life, the life of art" is the core value of this project, aiming to create an artistic club.

Different from the 2D approach of drawing, the sculpture is more stereoscopic and the sculptors complete the works from three-dimensional angle. Designer of this project is trying to realize the design of the space through sculpture, which adopts 3D modeling instead of 2D model. The whole design features the lines consisted of dots and a surface consisted of lines, then a three-dimensional space consisted of surfaces in different angles. Every line and every surface is something which is carefully carved by the sculptor and the final interior space design itself is an artwork.

Through geometric lines and lines of light, combined with marble mosaic grain and log color, a fashionable and leisure sense is highlighted in the lobby space. In the coffee area, the LED lamps emit soft light, thus creating a unique, elegant and casual atmosphere. The staircase is a piece of sculpture, and its irregular lines enrich the layering sense. With waves as the main line, the swimming pool shows the layers of the ceiling. Fitness zone contains a gym, dance studio, billiard room, table tennis room and other functional areas. In the basketball court, the walls and the ceiling form a graceful curve, and with a combination of LED lights, the whole space produces wavy layers.

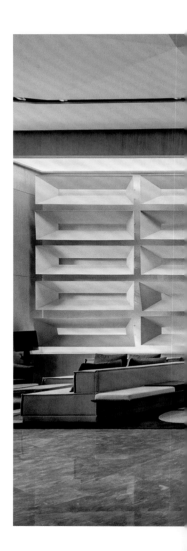

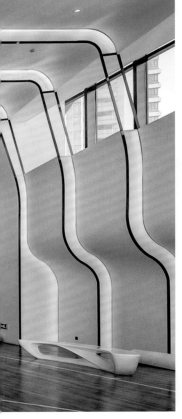

NOMINATION FOR THE BEST CLUB DESIGN

最佳会所设计奖提名奖

季青涛（中国重庆）
Ji Qingtao (Chongqing, China)

获奖项目/Winning Project

长滩壹号餐茶会所
Club in Long Beach Mansion

设计说明/ Design Illustration

2015年底，融创集团在成都的别墅项目长滩壹号的销售中心需要改造升级，主要目的是丰富区域内的生活体验，增加若干私享空间，而本案呈现的是其中的一小部分相对独立的餐茶分享区，这也决定了会所特有的自身空间气质的定位。

20世纪初，建筑先驱们提出了"Less is more"的生活美学，如今依然盛行，只不过有不同的诠释。之于本案，设计师巧妙地将这种诗意栖居的美学移植到会所空间中，让东方的"静"与西方的"净"加以结合，赋简约的精致以静谧的情绪，在多与少、黑与白之间，演绎着不同的人生哲学。

会所内的空间架构上深谙东方之气韵，于平淡中有惊喜，于回归中有人性温暖。设计师试图为使用者提供一处宁静、舒适的体验场所，让人安静地融于这自然中。

大面积木作，水墨画的清渺，古灯的空灵，梅花的清雅，一系列中式元素恰如其分，使得这方空间瞬间有了底蕴，也有了意境，于简约之中散发出浓烈的中国禅意情怀。同时为迎合人居环境低碳环保的理念，本案中严格控制材料的种类，原木、乳胶漆的重复使用，带入老树、马头墙、花窗、梅花等传统印象，以温婉柔情映射左右；回廊村落，峰回路转，烘托出一派"心静、人舒"的惬意。

At the end of 2015, Sunac China Holdings Limited required to remodel the sales center of its villa project Long Beach Mansion in Chengdu. The main purpose was to enrich living experience and add several private spaces, and this project is one of the small individual areas, now it's being used for dining and tea drinking. All these factors define the temperament of the space. At the beginning of last century, pioneers of architecture put forward the life aesthetics of "Less is more". It still prevails in current days, but has different explanations. In this project, the designer cleverly transplanted this aesthetics of poetic dwelling into the club, combined the Chinese "quietness" with Western "clearness", endowed simple exquisiteness with tranquility, thus interpreting various philosophies of life between more and less, black and white.

The oriental charm is embodied very well in the spatial layout, you can find surprises in simplicity and feel the warmth of human nature. The designer tried to provide a quiet and comfortable place which enables users to be a part of the nature.

Chinese style elements such as carpentries, ink and wash paintings, antique lamps and elegant plum blossoms are perfectly arranged, together they create a strong Chinese Zen style accent. Meanwhile, in order to cater to the low carbon and environment−friendly living philosophy, the designer strictly controlled the materials, for example, he repeatedly used raw wood and emulsion varnish, and introduced traditional impressions of old tree, ma tau wall (a classical wall of traditional architectures in Jiangnan areas of China), lattice window and plum blossom, all these add gentility and tenderness into the club. It's like a winding path in the village, along which you can see different sceneries that make you feel calm and comfortable.

NOMINATION FOR THE BEST CLUB DESIGN

最佳会所设计奖提名奖

河南贝铭设计装饰工程有限公司
（中国郑州）
Booming Design Construction Associates
(Zhengzhou, China)

. . . .

获奖项目/Winning Project

郑州润居茶水间
Zhengzhou Runju Tea House

设计说明/ Design Illustration

润居是以宋风为主题的茶文化空间，将宋人的儒雅自由、包容、谦和的精神特质及对琴、棋、书、画、诗、酒、花香、茶等生活之大美以空间作为载体来诠释。

项目以书院气息作为空间气质轴线，借中原特有的钧瓷"入窑一色，出窑万彩"的艺术特点，用深度考量后的中轴对称来凸显空间的空灵、淡泊、雅致而悠远的氛围。

空间中开阔而舒适的书库长廊能使人迅速卸下门外的疲惫，风化的树、斑驳的船，在大尺度的泼墨山水映照下并不突兀，宋代典雅平正，不失雕琢，质朴清雅及宋文化主导的简雅、隽永、和祥，在不显山不露水中一一铺陈开来。无论是场景的还原，还是书画的再现，美得浑然一体。在空间设计的时候刻意在自然的规则与不规则之间形成一种对话，让传统文化与现代时尚友好共融，把自然和仪式感在这个宽大的空间里很好地呈现出来。

润居这个项目更多程度上做的不是空间，而是内心。希望通过仪式感营造既古典，又相对自在、自由的空间。无论是寄情山水还是和大自然的其他交流，让进来这里的人能释怀、轻松。

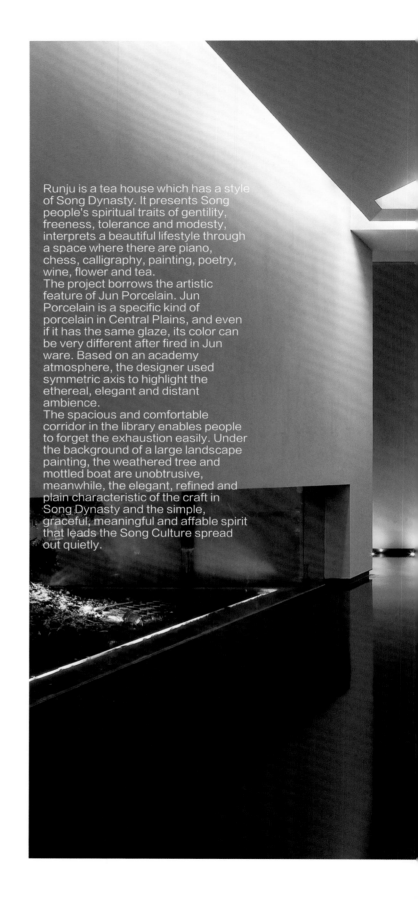

Runju is a tea house which has a style of Song Dynasty. It presents Song people's spiritual traits of gentility, freeness, tolerance and modesty, interprets a beautiful lifestyle through a space where there are piano, chess, calligraphy, painting, poetry, wine, flower and tea.
The project borrows the artistic feature of Jun Porcelain. Jun Porcelain is a specific kind of porcelain in Central Plains, and even if it has the same glaze, its color can be very different after fired in Jun ware. Based on an academy atmosphere, the designer used symmetric axis to highlight the ethereal, elegant and distant ambience.
The spacious and comfortable corridor in the library enables people to forget the exhaustion easily. Under the background of a large landscape painting, the weathered tree and mottled boat are unobtrusive, meanwhile, the elegant, refined and plain characteristic of the craft in Song Dynasty and the simple, graceful, meaningful and affable spirit that leads the Song Culture spread out quietly.

From the recovery of scene to the reappearance of paintings and calligraphies, all are beautifully blended in the space. The designer deliberately established a dialogue between regular and irregular items, integrated traditional culture and modern fashion in a harmoniously way, thus presenting a natural sense and spatial ritual. This project is more about the inner heart instead of space. The designer wanted to create a classical, but relatively free and comfortable space by spatial ritual. You can attach your feelings to the landscape or communicate with the nature, and either way can release your mind and make you feel relaxed.

370

The
Best
Villa
Design Award
最佳别墅设计奖

艾特奖

最佳别墅设计奖
THE BEST VILLA DESIGN AWARD

INTERNATIONAL
SPACE DESIGN
AWARD

CHANG
ARCHITECTS

获奖者/Winner
CHANG Architects
(新加坡/Singapore)
获奖项目/Winning Project
康沃尔花园/Cornwall Gardens

获奖项目/Winning Project

康沃尔花园
Cornwall Gardens

设计说明/ Design Illustration

这座别墅是为四代人建造的。客户计划他父母退休后可以在这里养老，一个"开放的家，清凉的热带天堂"，也能鼓励后辈们长大后也在这里安家。

设计理念遵循自然与人的神人关系，家庭与自然共享呼吸空间。植物、水体、居住空间都合而为一。别墅提供日照、自然通风和被动式降温，也提供可促进整体福祉的生态友好的环境。

在门厅处，一面被保留的渗水的墙被转化为一个带瀑布的绿色庭院。客人来访时可看到热带雨林植被和梯级水流声，这种来自自然的致意在各方面都是宜人的。

在处理已存在的地形时，为增加建筑密度而使用的阶梯被用作种植热带水果树的花槽，这样能冷却周围的气温，同时与室内隔离开来。在平面图上，别墅在邻里之间就像一片绿洲，景观层和阶梯植被形成了生物池。这是雨水汇集并循环灌溉的地方。

种植着多种类植被的阳台标识出房间的外围，并且能俯瞰中心水池。连续的阳台形成一座花槽桥，在里面种植鸡蛋果，蔓延的藤蔓可以提供荫蔽，以及遮挡邻里的私人屏障。

在这高科技的时代，这栋别墅展现了精致的建筑手艺。建筑设计师把原本视为废物的材料，巧妙地另类运用，打造出全新的感观及视觉体验。它打破了一代受欢迎的欧式风格，以简朴的材料融合大自然的魅力，摇身变成一栋有高贵气质的别墅。这优美的意境，让住户犹如住在一幅绝妙美丽的风景画中。

这座别墅已经成为家人和朋友受欢迎的聚会场所，也吸引了各种生物——蜜蜂、蝴蝶和松鼠。

在自然中生活，并持续与自然接触，这座别墅是避开拥挤城市生活的私室，使得灵魂与自然重新联结。它也反映了客户和咨询建造团队之间的紧密合作，从而使这座当代热带别墅项目得以实现。

This house is for four generations. The client had planned for their parents' retirement, an 'open home, a cool tropical paradise for the family', encouraging their children to 'raise their families here when they grow up'. Designed with an I-Thou relationship with nature, the family and nature share the same breathing space. Plants, water bodies, and living spaces are

integrated as one. The setting provides daylighting, natural ventilation, and passive cooling. It offers an ecological–friendly environment that promotes general wellness for all.

At the foyer, an old retaining wall with a history of leakage has been transformed into a green courtyard with waterfall feature. Visitors are now greeted with tropical rainforest plants and the sounds of cascading water, which can be enjoyed from all levels.

Working with the existing terrain, built–ups that contributed to the site coverage are utilised as planters for tropical fruit trees, to cool ambient temperature, and to insulate the interiors. On plan, the house is a green oasis amongst the neighbourhood, of landscape decks and cascading planters framing the bio pool and ponds. These are the catchment areas for rainwater harvesting, to be recycled for irrigation.

Planting verandahs of varying plant species line the peripheries of the rooms and overlook the central pool. This continues to form a planter bridge of passion fruits, where climbing vines provide sun–shade from the setting sun, and a privacy screen for the neighbours.

Craft is celebrated here. Salvaged materials and objects form part of the schedule of finishes. It suggests alternative palette, tactile and experiential qualities of housing, in physical and spiritual sense.

This house has become the popular gathering place for the extended families and friends, and it has also attracted a host of biodiversity – from bees, butterflies to squirrels.

By living with and constantly in touch with nature, this house is a sanctum from the hustle of urban–living, reconnecting one's soul with nature. It also reflects the great collaboration between the client and the team of consultants and builders, in making this contemporary tropical setting possible.

获奖评语

这个别墅项目用典型的和最简洁的元素在建筑中定义出了一个几何空间，但采用了一种全然不同的方式诠释所有的元素，从而将元素与环境、光、物质、水、绿化紧密结合起来。

This is the villa which the project remains a building in architecture, defining the geometry with typical and minimum elements in architecture,but interpreting all these elements in a totally different way, creating a strong relationship with the environment with all the elements, light, material, water, and greening.

NOMINATION FOR THE BEST VILLA DESIGN

最佳别墅设计奖提名奖

Greg Gong (澳大利亚/Australia)

获奖项目/Winning Project

五号别墅
Villa No. 5

设计说明/ Design Illustration

项目为位于绿树成荫的墨尔本郊区的一栋私人别墅。

别墅所在区域为遗产保护区，区域内场地建筑有高度限制。场地面积为808平方米，入口处有一段轻缓的斜坡。街道在场地南面，别墅也是朝南的——庭院区域可以被阳光照射。然而斜坡制造了一个严重的问题：别墅南部与街道处于同一水平线上，而北部几乎高了一个级别。建筑师的解决方案是将场地整体移平，与街道齐平。别墅整体面积为1200平方米，分布在三层楼。上层是私人区域，有主卧室、书房和次卧。首层是家庭区，分为南北两个主空间，各自与庭院很好地联结。南边是日式枯山水庭院，展示一种冷静和不断变化的景观。北边庭院通往家庭区，为别墅的室外空间，庭院周边是水池的倒影。8米高的围墙使得庭院像是一个没有屋顶的房间，这成为住户休闲活动的中心区域。家庭区连接着一个开放式的西式厨房，西式厨房后面是一个隐匿的中式厨房，这为住户提供两种不同的烹饪选择，以满足住户需求。入口区域是项目的亮点。这里被设计成双层空间，以提供天气防护，同时延展进别墅中心，联结了正式居住区和休闲活动区，楼上和楼下，客人接待区和服务通道。一条长板凳作为亮点提供了座位，也联结了楼梯井的天窗和楼上。负一层通过举车器可容纳5辆车。举车器在工作时车主不必下车。负一层有健身房和冲凉室，还有吧台和待客室。客人可从外面庭院直接进入负一层而不用穿过房屋。还有一个可调节温度的酒窖，完全线型的木柜可储藏4000多瓶红酒。

建筑师为设计这座别墅而在世界范围内寻找材料。标准长度的木地板是从德国山林区进口在丹麦制造。从木材被砍伐到运至澳大利亚需要10个月时间。这种特殊地板有15米长，被隐秘修复过。木地板下也有采暖。石板是为了项目特地在巴西制造的。所有玻璃门是在瑞士定制的。所有细木制品和家具是在澳大利亚和意大利定制的。

景观是由日本景观设计师Moto设计的，他家是景观设计世家，他是第三代传人，从小和祖父学习景观设计。

特别要指出的是，太阳下山后项目的照明系统将使别墅转变成另一种完全不同的气氛。

This is a private house built in leafy suburb of Melbourne.

It is located in a heritage protection area where there is a building height control over the site. The site area is 808 sqm2 with a gentle slope up from the entry area. The street is on the south side of the site which gives the preferred orientation to the site - leaving the northern (sunny side to the private garden area. However, the slope site creates a serious problem, this house has been set to sit on the street level which has to be about one level down from its northern neighbour. The architect heads on with the problem by bringing the site down to street level to make it works with the height control. The house is about 1200sqm spread in three levels. The upper level is the private realm with master bedroom and a study, as well as the second bedroom. Ground floor is the family area where two major spaces are allocated at north and south side with distinctive gardens well connected to each of them. The south side is a Japanese style dry garden which provides a calm and ever changing landscape. The northern side which connected to the family room, a paved courtyard provides a family outdoor space with a reflective pool surrounded the side. The 8 metres high garden wall makes the private garden just like a large private room without a roof. It becomes the focus point for the informal living area. The family room is joined with an open western kitchen while a fully concealed Chinese kitchen just behind the western one to provide the flexibility to have both style of cooking to meet the family needs.

The entry area is arguably the highlight of the project. It has a double volume space to provide weather protection extended into the centre of the house to create a wow factor to link formal/informal living area, upstairs and downstairs. Guest-family access/service access. A long bench acts as a focus point to offer sitting, as well as link the skylit stairwell to upstairs.

The basement accommodates 5 cars accessed by car-lift. This is the first-of-its-kind passenger car lift which the driver in the car does not required to get out of the car while it goes up or down. The basement has a gym with shower and a bar and also a guest room which can access from the sunken garden without going through the house itself. A climate controlled wine cellar fully lined with timber can storage over 4,000 bottles of wine.

The architect has source the material for the house over the whole world. The full length plank floor boards are made in demark with the material from black forest in Germany. It takes 10 months from harvest the timber to have the floor boards delivered to Australia. This particular floor boards is as long as 15m with concealed fixing. There is also hydraulic heating underneath the timber floor. The stone slabs are specially made for the project in Brazil. All the glass doors are custom-made in Switzerland. All the joinery and furniture

are custom-made in Australia or in Italy.

The landscape was designed and built by Japanese landscaper Moto who is the third generation of landscaper who hones his skill from very young age with his grandfather.

The lighting of the project needs special mention that transform the house into different mood when the sun goes down.

NOMINATION FOR THE BEST VILLA DESIGN

最佳别墅设计奖提名奖

谢佳妏（中国台湾）
Xie Jiawen (Taiwan, China)

获奖项目/Winning Project

中体西用的新东方艺术
The New Oriental Art in Westernized Chinese Style

设计说明/ Design Illustration

将本案的设计概念总结为：天圆地方中体西用，虚实交错明昶宽阔。从比例、均衡、行气三大方向落实设计概念，自东方经典中提炼出意义性图腾元素，解构风格本质，用实境与意境穿插，重新组构空间，渲染出空间的意义与特征，表现空间天圆地方中体西用的强烈主观，将巨大的概括力和极度简洁融合，强化美感和思想产生强烈的感染力。以现代工艺结合东方人文艺术为主轴，不仅架构出整体画面的美感也产生电影情节般的空间张力，将西方Lounge的低奢，剪接置入东方传统生活意境里，创造无限延伸的新奢华空间。

The concept of design in this project can be summarized as round sky and flat ground; westernized Chinese style; the staggered virtual reality; brightness and wideness. The design concepts are materialized in three different aspects such as ratio, equilibrium, and correspondence relationship. In the meantime, the meaningful totem elements are refined from oriental classic that deconstruct the essence of style. Furthermore, the meaning and characteristic of the space are revealed by intersecting real and artistic conceptions that reorganize space. The strong subjectivity, which is expressed by the ideas of round sky and flat ground and westernized Chinese style, fuses the gigantic generalization and extreme simplicity, and strengthens the strong infection generated by aesthetics and ideas. On the basis of modern craft and oriental humanity art, not only the integral aesthetics of image generate a kind of spatial tension like a movie scene, but also the design embeds the low-profile luxury of western lounge into oriental traditional life artistic conception, and creates unlimited extension in extravagance of mind.

NOMINATION FOR THE BEST VILLA DESIGN

最佳别墅设计奖提名奖

张智琳（中国台湾）
Zhang Zhilin (Taiwan, China)

获奖项目/Winning Project

四季别墅
Four Season House

设计说明/ Design Illustration

这个绿色房子位于云林县，云林是纯朴的农村城市。对于基地的第一印象是被农田、竹林和乡村道路所包围。在此基础上有几棵原始树种的存在。我们保留现有的树木，并且将其作为建筑配置及景观规划的主轴，因为我们相信自然物种是珍贵的，建筑应该是尊重并且顺应环境，所以我们尽量减少对自然的影响。因此，建筑设计的规划上，使建筑与大树处于相对的位置上，无论是天际线或是方位配置，都是相互对应并且融入的。

这座房子有5个家庭成员使用。在一楼有停车空间、客厅、厨房、休憩室和一间卧室。我们的视线可以一路从客厅到餐厅及厨房，将楼梯作为建筑中心，串联起所有空间与家庭活动，没有实墙的固定隔阂让空间更为流动顺畅，也使其充满更多想象与期待。

我们可以透过落地窗，看见楼梯旁边的庭院景色，在楼梯上下空间里，我们也可以感受到时间与季节所带来的变化，深切地感受到环境与生活同步。当进入这座房子，我们可以看到沉浸在阳光下的大树，随着微风飘动的树影。从日出到日落，我们可以感受到不同的光影变化，就算在家里，也能清楚感受到季节与时间的变化。

创造似乎不存在的边界，强调室内及户外的联结。共存是我们想遵循的理念。

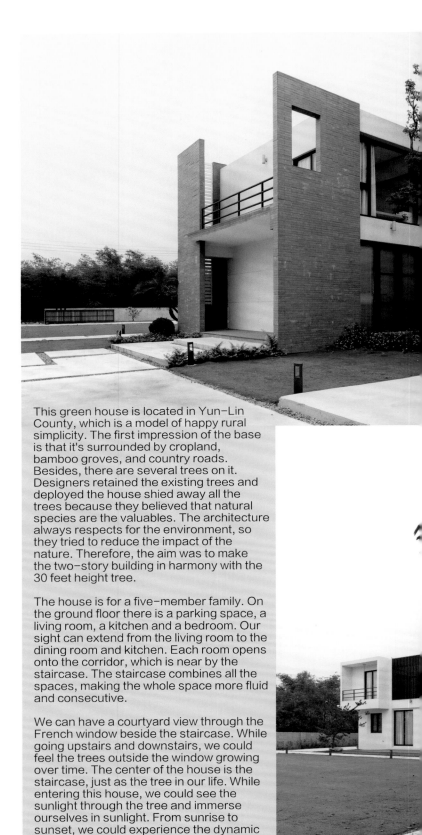

This green house is located in Yun-Lin County, which is a model of happy rural simplicity. The first impression of the base is that it's surrounded by cropland, bamboo groves, and country roads. Besides, there are several trees on it. Designers retained the existing trees and deployed the house shied away all the trees because they believed that natural species are the valuables. The architecture always respects for the environment, so they tried to reduce the impact of the nature. Therefore, the aim was to make the two-story building in harmony with the 30 feet height tree.

The house is for a five-member family. On the ground floor there is a parking space, a living room, a kitchen and a bedroom. Our sight can extend from the living room to the dining room and kitchen. Each room opens onto the corridor, which is near by the staircase. The staircase combines all the spaces, making the whole space more fluid and consecutive.

We can have a courtyard view through the French window beside the staircase. While going upstairs and downstairs, we could feel the trees outside the window growing over time. The center of the house is the staircase, just as the tree in our life. While entering this house, we could see the sunlight through the tree and immerse ourselves in sunlight. From sunrise to sunset, we could experience the dynamic light and shadow inside the house. Along

with the changes
of seasons, the
view changes, too.

The interior space
and the outdoor
area are strongly
connected, so it
seems that there is
no boundary here.

Coexistence is the
main idea the
designers follow.

NOMINATION FOR THE BEST VILLA DESIGN

最佳别墅设计奖提名奖

徐友贤（中国上海）
Robert Su (Shanghai, China)

获奖项目/Winning Project

山水壹号
Heaven in the Sky Villa

设计说明/ Design Illustration

本项目"山水壹号"位于南通开发区，西邻君山，通往长江的天然河道沿小区中央蜿蜒流过，为望山瞰水的低密度高端别墅项目。
设计师在空间设计上采用了现代简约的表现手法，优化平面，使整个空间通透明亮，连贯融合，动线流畅，整个立面既整体干净，又对比鲜明，使整个空间色彩明快，简洁大气，低调内敛又不失精致高雅，同时配以线条简洁、色彩大方、搭配自然的家具饰品，重点强调低碳环保的装饰材料的使用，完美彰显了品味独特的国际设计风范。
别墅在功能布局上，设计师充分利用并优化原始平面，在上叠别墅的第三层设有入口玄关、客厅、餐厅、中西厨房、卫生间等；在第四层设有次入口玄关、书房、两个次卧室、衣帽间、卫生间、阳台等；第五层是主卧总统套房，设有主卧室、主卫生间及衣帽间、南北露台以及阁楼多功能室。
材质上设计师采用了大理石、地砖、木地板、墙纸、木饰面、玻璃、卡其色皮、不锈钢、人造石等，并严格把关所使用材料具有低碳环保性能。
在灯光的设计上，设计师尽量避免使用大面积刺眼的光源造成光污染，而采用暗藏式的射灯与落地灯、台灯相结合，使房间的氛围自然温馨，亲切优雅。

This is a low-density high-end villa that is located in Nantong Development Zone. It's near Junshan Island, and a natural waterway which leads to Yangtze River meanders through the center of the community.
The designer adopted a modern simple method into the space. After optimizing the plane, the whole space becomes clear and bright with a smooth circulation. The clean elevations form a striking contrast among each other, and they make the place not only lively, simple and low-key, but also elaborate and elegant.
The lines are concise, and with a match of natural furniture which emphasize on low carbon and eco-friendly materials, they manifest a special international taste in a perfect way.
In terms of the functional layout, the designer arranged entrance foyer, living room, dining room, Chinese and Western kitchens as well as bathroom on the third floor. The fourth floor also has an entrance foyer, and there are other spaces like study, two bedrooms, cloakroom, bathroom, balcony, etc. On the fifth floor, there is a presidential suite which contains a main bedroom, bathroom, cloakroom, south-north terrace and a multifunctional room in the attic.
The designer used lots of materials, including marbles, tiles, wood floors, wallpaper, wood veneers, glasses, khaki furs, stainless steels and artificial stones, all are strictly controlled to make sure they are low carbon and environmentally friendly.To avoid light pollution caused by large area of dazzling lights, the designer applied hidden lamps and floor lamps, thus creating a natural, warm, intimate and elegant atmosphere.

394

The
Best
Flat
Design Award
最佳公寓设计奖

艾特奖

最佳公寓设计奖
THE BEST FLAT DESIGN
AWARD

IDEA TOPS

INTERNATIONAL
SPACE DESIGN
AWARD

获奖者/Winner

洪文谅（中国台湾）

Hung Wenliang (Taiwan, China)

获奖项目/Winning Project

693号·一/No. 693·One

获奖项目/Winning Project

693号·一
No. 693·One

设计说明/ Design Illustration

当人的本体进入到空间实相，感受生命存在
与宁静。将设计润入自然且与环境交融，于
空间度量中，透过材质或陈设的脉络与文
化，围塑出看得见却听不到的声音，让心情
轻松，想象飞扬。
在石与木的脉络对话中、光晕开的面积里，
桌椅之间的手制精工文化悄悄拢入，不仅成
就生活初妆，亦在有形与无形的轨迹下，植
入纯粹、悠缓的静谧旨趣。

Life's existence and tranquility are
felt immediately when a human is
added to the picture. When a design
prioritizes nature and is well blended
into the environment, the context and
culture displayed by the materials and
the layout generate a voice that can
only be seen and not heard, making
one feel relaxed and joyous.
Craftsmanship and refined taste of the
chair and table are carefully nested in
the contextual dialogue between rocks
and wood and can be discovered
when light is shone onto the surface.
Purity and tranquility are inserted,
through tangible and intangible
expressions, in life's very basic
comforts.

获奖评语

它每一个空间都用得非常到位，而且有很多细节，这些细节组合起来形成了一个很好的画面。对细节的追求在整个空间里面非常成熟。

Every space is used very properly with a lot of details. These details together form a very good picture. The pursuit of details is very mature in the entire space.

NOMINATION FOR THE BEST FLAT DESIGN

最佳公寓设计奖提名奖

洪文谅（中国台湾）
Hung Wenliang (Taiwan, China)

获奖项目/Winning Project

58号·念
No. 58·Receive

设计说明/ Design Illustration

以纯粹而安静的本质，形构出居宅丰富的生命力，以线面的连贯维系持续性，借由筛光的动作，形成立面的延续主题，让阳光的温度围拢全室，使视觉在游走的同时，衍生出让居住者体会"有趣"的强大感染力，可以领略空间中的和谐的律动，材质发展成主要轴线，形式上具备简约朴素、宁静自然为范的"侘寂"概念逐渐明朗。

材质返璞归真的质蕴，墙与木之间的纹理，依循着比例、光影及颜色透着自然的旨趣，经过融合或隐匿的方式表现，让场域获得更大的放松。

Simplicity and serenity bring vitality to the house. Consistent lines create a sense of coherence and natural lighting brings out a thematic structure, warming the room. All it takes is a look around the room for residents to be inspired by the "funness" and harmonious vibration. By letting materials take the lead, the "wabi-sabi" philosophy, which emphasizes simplicity and nature, takes manifestation.

The simplicity of materials and the surface temperature of the wall and wood both reflect naturalness through their proportion, lighting, shade and color. The presentation, in the form of fusion and obscurity, enables the space to function better as a relaxing area.

NOMINATION FOR THE BEST FLAT DESIGN

最佳公寓设计奖提名奖

Daniel Bismut（法国）、凤丹、郭晓阳（中国上海）

Daniel Bismut (France), Feng Dan,
Guo Xiaoyang (Shanghai, China)

获奖项目/Winning Project

经典与永恒现代公寓设计
Classic and Eternal Modern Apartment Design

设计说明/ Design Illustration

永恒不是不变的，而是适应万变的，把传统过渡到现代，让美延续，让经典成 为永恒。本案位于巴黎16区一幢19世纪的传统石质古建筑中。是由酒店和居民楼组成的一个住宅区。在本案设计中，本着对传统的尊重，保留了这个公寓的古老性格，同时引入现代元素，成就了一个以经典永恒为美的现代设计。

细节决定成败。无论颜色还是材质，讲究的不是杂乱的堆叠，炫酷的视觉效果，而是一气呵成，恰到好处。住宅空间首要的是住的需求，家的感觉。一个好的设计要有灵魂、有内涵、有气质和性格。这也是我们BISMUT团队对每一个方案的态度与追求。

在入口的设计中我们保留了欧式古典的墙面雕塑结构，并以水泥做饰面处理。暖气功能被隐藏在镜面不锈钢的造型装饰面板中，最大程度地凸显了陈列品和实用相结合的艺术品位和价值。

视线穿过门洞，整个空间通透优雅、层次丰富、一气呵成，体现了欧式设计的别样禅意。客厅空间中，一面陈列背景墙体现了多种材质的精致搭配和点到为止的色彩运用。哑光

熏黑鸡翅木所装饰的天花板与黑色哑光漆处理后的经典凡尔赛实木地板交相呼应，并致力于平衡经典与现代元素、材质、色彩的完美衔接与对比。此外本案中的一部分家具，如书桌、茶几、边几、书架、餐桌等也是由本案设计师做配套设计。

镜面不锈钢餐桌的设计是整个餐厅的灵魂，经典的欧式线条，极简的现代天花板和与之配合的黑色窗框，都跳跃在桌面的反射中。经典的棋盘格大理石地板被桌腿的曲面像哈哈镜般扭曲变了形，让整个空间庄重典雅又不失活泼。

主卧空间中，银色织物软包、熏黑鸡翅木和青铜的置物格镶嵌在青铜的框架中，好像一幅画中画，使床头与床融为一体，实用美观的同时还隐藏了宽厚的梁，并巧妙利用了梁底空间。

主卧浴室是由手工切割的熔融玻璃装饰，它由铁蓝色、紫色马赛克和镀24k金的马赛克组成，搭配华美的意大利天然大理石。

Eternity is not something that never change, it is adaptable. With transition from traditional to modern style, let beauty continues and let classic become eternal. This project is located in a 19 century traditional stone ancient building in the 16th district of Paris. It is a residential area composed of hotel and residential building. In this project, in respect of tradition, we kept the ancient character of this apartment, at the same

time adding modern elements, thus achieving a modern design which beauty is classic and eternal.

Details determine success or failure. In terms of color or material, what matters is not random stacking or cool visual effects, but to finish in a smooth way and everything is in its perfect place. The primary requirement of residential space is living, a feeling of home. A good design has soul, content, temperament and character. This is also the attitude and pursuit of our BISMUT team.

In entrance design we kept European classic wall sculpture structure and put cement on the surface. Heating function is hid in the stainless steel decoration board with mirror surface, which at the most extent expresses the art taste and value of combining display artworks and utility.

Seeing from the door hole, the whole space is transparent and elegant with fruitful layers, which conveys a special Zen style of European design.

In living room space, a display background wall expresses delicate matching of multiple materials and light touching of color. Matte dark wenge ceiling interfaces with black matte paint classic Versailles solid wood flooring. And we also devote to balance the perfect connection and comparison of the classic and eternal elements, materials and colors.

The design of stainless steel dining-table with mirror surface is the soul of the whole dining room. Sky of classic European lines, minimal modern ceiling and black window frame are jumping in the reflection of the desk. Classic chessboard grid marble floor is twisted by the curve surface of table leg, like a distorting mirror, which makes the whole space solemn and elegant at the same time lively.

In master bedroom space, silver fabric upholstery, matte dark wenge and bronze glove box are inlaid inside the bronze frame, like a picture in picture, which combines bed head and bed into one part, properly hiding the wide and thick beam at the same time being practical and beautiful, and cleverly using the bottom space of the beam. Master bedroom bathroom is decorated by manual cutting of fused glass, which is composed of iron blue, purple mosaic and 24k gold plated mosaic, with ornate Italian natural marble.

NOMINATION FOR THE BEST FLAT DESIGN

最佳公寓设计奖提名奖

邵荣（中国青岛）
Shao Rong (Qingdao, China)

获奖项目/Winning Project

康桥首府
Cambridge House

设计说明/ Design Illustration

本案例中，从顶面到地面都是使用最天然、朴实的材料，隔离了城市的喧嚣。淡木色的地板，散发出原生态的大自然气息。温润的木色与简单的白色相搭配，营造出舒适温馨的居家环境。圆形的藤编式坐垫日式风格浓厚，木质的收纳柜可以摆放小植物等装饰品，使得空间更富情调。角落处的小收纳柜，可以放上家中常用的物品，方便取拿。白色的花瓶上，插上一束花，仿佛散发出淡淡的清香。绿色的植物自然气息浓厚。正方形的木质茶几，木纹清晰，给人文化认知与不失素雅。一套白色的沙发，简单质朴，体现出主人悠闲的生活情调。厨房木质边框的玻璃门，尽显通透之色。书房淡绿色的榻榻米，吹来了淡雅的恬淡气息，搭配炕桌，简单自然，一株干花，高贵优雅。嵌入式的木质收纳柜，不仅非常节省空间，还可以放上家中的大件杂物等，使得小家更加干净整洁。

From the ceiling to the floor, all the materials used in this project are natural and plain. The light wooden floors exude a natural atmosphere which keeps off the noisy feeling of the city. The mild wood color is perfectly matched with simplewhite, creating a comfortable and warm living environment. The round rattan cushions show a touch of Japanese style, and you can make the space more appealing by displaying decorative items on the wooden cabinet, such as small plants. Besides, the cabinet at the corner is very convenient for you to put and get commonly used things. Flowers in the white vase are sending forth a delicate fragrance, and the green plants add a strong natural ambience. You can see wood grains clearly on the square tea table, which may remind you of cultural awareness. A set of simple white sofas reveals the host's leisurely style of life. Glasses were embedded into wooden flames of the kitchen door to make a transparent space. A simple and natural mix of light green tatami, Kang table and dried flowers achieves a graceful atmosphere in the study. The built-in cabinet saves lots of space and is available for large items, so that your house can be cleaner and tidier.

NOMINATION FOR THE BEST FLAT DESIGN
最佳公寓设计奖提名奖

广州名艺佳装饰设计有限公司
（中国广州）
James Liang & Associates Limited
(Guangzhou, China)

获奖项目/Winning Project

保利金融大都汇——米墅
Poly Tal To Wun—Mels Villa

设计说明/ Design Illustration

为了能突破传统公寓与复式公寓给我们带来的不足，我们决定作出大胆的尝试，使得传统平层公寓与复式公寓两者合二为一，将仅有3.5米高、35平方米的平层公寓转变成五脏俱全的小型复式公寓，令您用只能买到普通平层公寓的钱就能享受到一间面面俱到的复式公寓——米墅。

米墅，通过利用空间的错落，打破传统的公寓空间限制性，模糊传统的复式公寓对层高与楼梯的界限，极大地提升了储藏的空间，并将空间分区更合理、更功能化。米墅的设计，最大的亮点就是拥有巨大的储藏空间，利用所有可以利用的空间转化为储藏间。

米墅的设计概念，源于我们针对传统的平层公寓与复式公寓弊端，利用+、−、×、÷四元数来进行拆解。"+"（对空间的叠加）"−"（减掉繁琐的累赘，简约生活方式的体现）"×"（有限的空间乘以无限的遐想）"÷"（3.5米除以2）=米墅

我们用有限的空间，为您创造出无限的可能，用极致追求的想象，实现"一户多变"的居住梦想，提高购买或租赁欲望。小空间大生活，用无限可能存放最好的青春年华！

To break through the shortcomings of traditional flat floor apartment and duplex apartment, we decide to make a bold attempt, to make traditional flat floor apartment combine with duplex apartment. We change the small flat apartment with only 3.5m height and 35 square meters into a duplex apartment fully equipped, therefore you only have to spend the price of a normal apartment to enjoy a fully equipped duplex apartment—Mels Villa.

Mels Villa, using the disorder of space to break through the limitation of traditional apartment space and blur the floor height and staircase limitations of traditional duplex apartment, which enlarges storage space hugely and divides the space partition more reasonably and functionally. The biggest bright spot of Mels Villa design is its huge storage space, transforming every usable space to storage room.

The design concept of Mels Villa origins from below idea: aiming at disadvantages of traditional flat floor apartment and duplex apartment, we use quaternion +, −, x, ÷ to disassemble it. "+": Spatial overlay; "−": Subtract cumbersome drag; "x": Limited space plus infinite imagination; "÷": 3.5m divided by 2. And the result is Mels Villa.

We use limited space to create infinite possibility for you, using our best imagination to realize the living dream of "one house with multiple forms" thus increasing buying or renting desire. Small space with big life, using infinite possibility to store the best youth!

418

P

The
Best
Show Flat
Design Award

最佳样板房设计奖

艾特奖

最佳样板房设计奖
THE BEST SHOW FLAT
DESIGN AWARD

INTERNATIONAL
SPACE DESIGN
AWARD

获奖者/Winner

深圳市昊泽空间设计有限公司（中国深圳）

Shenzhen Horizon Space Design (Shenzhen, China)

获奖项目/Winning Project

隐庐/Hidden House

获奖项目/Winning Project

隐庐
Hidden House

设计说明/ Design Illustration

都市蕞林的名利和欲望已经固化成一种执念，侵蚀我们每个人。迷失、焦虑变成一种无可回避的精神困局，人人都应在麻木与不屑中时刻警惕自己：我们是谁？我们在哪里？我们将要去何方？我们冀望以物质空间为载体，通过不同空间功能的思想性，用最本质的生活元素来唤醒人身体的各种触觉感官的敏感度，唤起自身的存在感，从而思考自己、思考人生、体悟生命真实的喜悦和智慧。

Fame, wealth and desire have become the obsessions which have a strong influence on us. It's unavoidable that we feel lost and anxious. Everyone should always ask questions like "Who are we?" "Where are we?" "Where should we go?" We hope to wake up the sensibility of our body by the most essential living elements, such as material space and different functions within the space. Therefore, we will feel the sense of presence, and think about ourselves, our life and realize the true happiness and wisdom.

获奖评语

把文化和情感的意念放在一个现代的样板房里面，很有内涵、很有感觉的空间融入这些现代的生活里面，非常成功。

Putting cultural and emotional concept into a modern show flat, and integrating space that full of content and feeling into this modern life, is very successful.

NOMINATION FOR THE BEST SHOW FLAT DESIGN

最佳样板房设计奖提名奖

深圳市盘石室内设计有限公司
（中国深圳）
Huge Rock Design (Shenzhen, China)

获奖项目/Winning Project

杭州·长龙领航90户型样板房
Hangzhou Changlong Pilot 90 Show Flat

设计说明/ Design Illustration

中国梦之《翔·梦》

少年时的梦想，常被认为是妄想，不被认可，甚至被嘲笑……但自己的梦想坚持了只有自己知道，我们做着自己喜欢的事情，独自走在路上乐此不疲。说来也巧，儿时对天空的向往如同一种寓言，命运将我们引向飞行。当年只在心中勾画的梦想，如今像新生的叶子一般，娇嫩又充满着生机，它已准备好挑战风霜雨露，迎着朝阳，抱着超越平凡的执着，圆梦高飞。勇敢放飞的梦想，它是超凡脱俗的，是无拘无束的，让我们凌空俯瞰大地，开启新的生命之旅。

当放飞翅膀翱翔在云端居高临下时，我们的视野感觉到的是无边际的"白"，所有的光、雨、大地都是通过云层的舒展进入我们的视野中。白茫茫的云端，永远是人类梦想飞翔的地方，那里清澈安宁、没有功利的羁绊、没有尘世的束缚，只需快活于自己的一片天。

《翔·梦》户型的设计初衷亦是如此。"非淡泊无以明志，非宁静无以致远"，虽无绚丽的彩虹，但它宁静而浩瀚，朴实无华中带着与众不同的浪漫。这里是"逐梦者"的飞翔乐巢，这里有他们美妙的梦想与渴望的世界。他们追求的不止于简单的飞翔，他们要享受着临空带来的乐趣，愿在心旷神怡中自由飞翔、透过云端俯瞰大地山河，飞得更高、飞得更远。

Chinese Dream of FLY•DREAM

The dreams we had when we were young are often being considered as delusions, without being admitted and even being laughed at...But only ourselves know that we have been sticking to our dreams all the time. We do what we love and walk alone in this road without feeling tired. Coincidentally, childhood yearning for the sky is like a fable, and we are led by the fate to fly. The dream that we had in our heart, now is like a new leaf, delicate and full of vitality. It now is ready to challenge all the difficulties, facing the sun and embracing persistence beyond ordinary, then fly! Bravely flying dream is free and beyond ordinary. Let us overlook the ground and open a new journey of life.

When flying in the cloud and overlooking the ground, the vision we have is infinite white. All light, rain and ground enters into our sight through the spreading of cloud. White clouds are always the place that human beings had dreamed of flying. That place is clear and quiet, and there are no fetters from utility, only to happily enjoy one's own world.

The original design intention of FLY · DREAM show flat is also like this. Like an old saying, one can only live simply to know and achieve his ambitions. Even though there is no gorgeous rainbow, it is quiet, huge and plain with special romance. Here is the happy home for dream chasers to fly, where they can find their wonderful dreams and desired world. Not only simply flying that they chase after, but also they enjoy the fun flying in the air, wishing to fly freely and happily, overlooking mountains and rivers through the clouds and flying higher and farther.

NOMINATION FOR THE BEST SHOW FLAT DESIGN

最佳样板房设计奖提名奖

唐列平（中国佛山）
Tang Lieping (Foshan, China)

获奖项目/Winning Project

筑象
Silence

设计说明/ Design Illustration

与自然对话

建筑的现象，即是人的现状，我们希望用一种新的、开放的方式重新去组合人的思维方式以及生活方式。如何将阳光与气流引入室内，是设计的基本，设计本应由这些自然元素去塑造，同时希望生活在里面的人，站在空间的任何角落，都可以很直接地感受这些，意味着，我们重新在室内做了一遍建筑。

安静是最终的目的，居住空间脱离了安静，容易缺失最本真的求知欲。这种安静，是精神的寄托，从而衍生新的生活习性。

空间，因为人的思想而产生区别，也因使用者的习性而作为基点，同质化严重的今天，是生活方式、认知的趋近所致！

我们尝试透过这种迷离不清的关系去寻找差异，这种谋求差异的过程即是设计的一部分，设计与不设计，更多的，是欲望的取舍。透过观察人的行为，想象行为之余的未知。在未知中，呈现沉默和安静，同时通过不同的角度去控制，达到人、空间、自然共同思考的方式。如此，我们称之为"筑象"。

The appearance of the building is the current situation of people. We hope to use a new, open way to recombine people's way of thinking and lifestyle. How to lead the sun and air into the room is the basis of the design. The design shall be shaped by these natural elements, and we hope that people live in it can feel it directly standing in any corner of the room, which means we've redesigned the interior.

Quietness is the ultimate goal. Living space without quietness is easy to lack its most genuine curiosity. However this quietness is the spiritual sustenance, which derives new living habits. Moving with the space is the basis of controlling the space temperature.

Space is different according to different people's thoughts, and is different because of the different space using habits of people. The serious homogenization of people's lives today is caused by the same cognitive and lifestyle approach of people!

We try to find the difference through this blurred relationship. The process of looking for the difference is part of the design. To design or not, what actually matters more is choosing to desire or not. We observe human behaviors to imagine the unknown left by the behaviors. In the unknown, the silence and quietness exist together. At the same time we try to control to reach a way of thinking together about people, space and nature through different point of views. Like so, we call it "Silence".

NOMINATION FOR THE BEST SHOW FLAT DESIGN

最佳样板房设计奖提名奖

磐锦室内装修设计有限公司
（中国台湾）
Pan Jin Space Design (Taiwan, China)

获奖项目/Winning Project

天绿
SKY GREEN

设计说明/ Design Illustration

如花般绽放 独具风格的生活况味

整体的空间设计，从人们最简单的生活需求发想，以及对设计的热情及专业，演绎出空间的多样风格，将对人的关怀、对空间的张力掌握，转化建构出一种生活态度与空间美学，让室内处处流露出恬静美好的氛围。为业主提供美学与实用相互呈现的美好，营造出细腻独特的美学空间。

由七种灰黑白原石切割堆栈出的几何渐层组合成的墙面语汇，并在其中几块加入了会发光的透光石，仿佛在黑夜中指引方向的星座低调地闪烁着，在简约空间中产生视觉的凝聚力量，进而营造出室内的层次分明，以及充满设计感与自在的休憩空间。

流露自然悠闲 沉浸于家的自在

在空间里运用不同的材质来堆栈层次上的立体质感，企图形塑一个轻透的空间器皿，即使位置在高楼层，却能将户外的绿意无落差地引进室内，摇曳的树影投射在手工染制的棉麻窗帘上，映出自然有机的倒影。让无机的空间，在一点一滴的设计下，渐渐由建筑转化成家，成为一个让心灵有所寄托与休憩的场域。

Blossom like flower
The unique style of life taste

The design of the whole space is thinking from the simplest life requirement of people, combining passionate and professional attitude toward design, therefore interpreting the space in a variety of styles. The design transforms the care for people and the control of space tension into a construction of life attitude and space aesthetics, so that the room is in quiet and beautiful atmosphere everywhere, providing a kind of beauty combining aesthetics and practicability for the client, and creating an exquisite and unique aesthetic space.

The wall vocabulary composed of geometry layers stacking by seven grey, black and white stones, with several emitting sunshine stones in it, are like constellation directs guidance in the dark night, twinkling quietly, which creates a visionary gathering power in this simple space, thus creating layers effect inside the room, and a rest place full of design sense and freedom.

Reveal nature sense and leisure
Immerse in the comfort of home

Different materials are used to stack three-dimensional quality layers in space, trying to create a light space vessel. Even though the project is at the high floor of the building, it can bring in the outside greenings with no gap. Swaying tree shadows projects on hand dyed cotton curtains with natural and organic reflections. With every bit of design, architecture in inorganic space could gradually be transformed into a home, a place for heart to rest on and relax.

NOMINATION FOR THE BEST SHOW FLAT DESIGN

最佳样板房设计奖提名奖

Minas Kosmidis (Architecture in Concept), YuQiang & Partners
(希腊 & 中国/Greece & China)

获奖项目/Winning Project

东莞 - 中国（原创）样板房
Dongguan - China (Origins) Show Flat

设计说明/ Design Illustration

项目是一幢新建住宅楼22层的一个250平方米公寓的室内设计。住宅楼是东莞市一个宏大居住房地产项目的一部分，这片居住区向深圳市延伸，因此楼盘给人的印象为深圳大都会的感觉。这个项目既是楼盘的样板房，也供潜在买家参观。

电梯到公寓入口之间有一个私人大堂，同时迎接主人和客人。大堂里也有足够空间可放置柜子和储物间。进入公寓后，首先是一个客厅，拥有各种子空间，比如用餐区、休息区和更重要的茶座。喝茶的习惯和中国生活方式密不可分，因为中国家庭内的每一段重要的哲学交谈都发生在茶桌旁。为了和用餐区分开，设计采用了一块金属框架的玻璃隔板，这块隔板融合了西方包豪斯运动设计原则和中国传统住宅的半透明玻璃板，在视觉上用餐区并未和客厅其他区域分开。餐桌旁是一个吧台，可以储存红酒和鸡尾酒，通过一道滑动木门与客厅相连。

中式厨房和烹饪方式需要额外的空间柜台和燃气锅炉。同样重要的是厨房被完全隔离，因为中式烹饪会造成很大的味道，不能影响公寓其他地方。最后，从公寓的两个阳台上，

可以俯瞰整个城市，也直接通往客厅。在室内部分，Minas Kosmidis事务所决定使用朴素的米色（大理石地板，墙上和天花板上的粗糙涂料），以创造一个适合画作摆放、面料图案、家具木材和滑动门的中性色彩。

所有公共使用区和储存区都被胡桃木饰表面的滑动门隔开，以便制造私人区域，不被外界视线打扰。所有公共使用区和储存区都被合理地设计，物材和面料选择米色基调，为住户提供一种舒适的感觉。小物件的摆放也创造出一种有趣的空间组合。一些画作被恰到好处地放置在每个房间，使得整个公寓有一种艺术馆的感觉。

This project is about the interior design of a 250 square meters apartment on the 22th floor of a newly built residential tower. The building is part of a much larger real estate residential project in the city of Dongguan, which due to its extensive development and expansion towards the city of Shenzhen in southeastern China, thus giving the impression of being part of Shenzhen's metropolitan area. The apartment was designed in order to serve as a show flat for the whole residential project, i.e. a newly built apartment that is decorated and furnished for prospective buyers to view.

The entrance in the apartment is possible through a private hall that leads from the elevator to the entrance door and serves as a welcoming hall for the apartment's owners as well as their guests. In the same hall there has also been left enough space for a small closet as well as a storage room. Entering the apartment, firstly, there is a living room with all of its common sub-spaces, such as the dining area, the lounge area and more importantly the tea corner, which is a habit absolutely inseparable with Chinese culture and its way of living.

Furthermore, every substantial and philosophical conversation in a Chinese family takes place around a tea table. To separate from the dining area, a glass panel with metal structure has been designed. It combines western design principles of the Bauhaus movement with semi-transparent panels of traditional Chinese

houses and doesn't actually isolate visually the dining area from the rest of the living room. Next to the dining table, the architects have created a room for the needs of wine storage and cocktail mixing which serves as a home bar and can be directly accessed by any side of the living room through a sliding wooden door.

Moreover, due to the particularities of Chinese cuisine and its way of cooking, there is a need for extra spacious counter-top in the kitchen and large gas hobs. What is equally important is the possibility of complete isolation of the kitchen because Chinese cooking can cause quite strong odors which must not spread to other areas of the apartment. Finally, the apartment's two balconies, which have panoramic view over the city, are directly accessible to the living room. In the interior shell of the house, Minas Kosmidis architects have decided to use materials of earthy, beige colors (marble slabs on the floor, coarse coatings on walls/ceilings) in order to create a neutral base for the promotion of the paintings, the patterns on fabrics and the wooden veneers on furniture and sliding doors.

All common use and storage spaces can be isolated by sliding doors covered with walnut veneer sheets in order to leave all vertical surfaces clear of any visual disturbance when desired. Both common use and private spaces in the house were appropriately designed to create a cozy feeling to the user; this was achieved through the use of earthy – beige shades of materials and fabrics, leaving the contrast of smaller elements to create an interesting composition in space; a balanced composition of paintings, which are wisely placed in each room, give the apartment a sense of art gallery.

442

The
Best
Display Art
Design Award

最佳陈设艺术设计奖

艾特奖

最佳陈设艺术设计奖

THE BEST DISPLAY ART DESIGN AWARD

INTERNATIONAL SPACE DESIGN AWARD

获奖者/Winner
王刚（中国北京）
Wang Gang (Beijing, China)

获奖项目/Winning Project
ROSE——首旅集团旗舰诺金酒店大堂装置艺术（2015）

获奖项目/Winning Project

ROSE

设计说明/ Design Illustration

2015年6月5日，中国第一奢华酒店品牌
"诺金"旗舰店——北京诺金酒店盛大开幕，
引起国内外的广泛关注。作为诺金酒店品牌
的名誉艺术顾问，著名艺术家曾梵志先生的
团队负责整个酒店的艺术品设置与选择。期
间，他们邀请数名艺术家与建筑师专门为北
京诺金酒店进行艺术创作。
ROSE处于一个比较昏暗的环境当中，它悬
挂于地下一层通往夹层的公共楼梯平台上
方。那ROSE核心处的一盏忽明忽暗的花芯
格外引人入胜，似乎触手可及又似乎深不可
测，纤细的杆件由新型环保复合材料——竹
钢制成，形成了令人难以理解的长细比，空
灵与韵味由此产生。
这件艺术品有着阿基米德式的几何逻辑，任
何一个角度看过去都有一种无限向内延伸的
吸入力量，也许这就是艺术家把这件装置命
名为"ROSE"的原因。
它同时令人惊叹地构成了中国传统哲学的
"无极"形态，这很容易让人联想到中国的
太极八卦图，但呈现出的气质却又是如此现
代而明朗。
夹层空间上楼梯两侧的月亮门恰好界定了
ROSE的空间范畴，方与圆在这里交相辉
映，让我们联想到"时光隧道"，也联想到
《盗梦空间》。所以，"最清晰的梦幻"也
许是对这件艺术品的最佳注解。

On June 5th 2015, the first luxury hotel brand in China "NUO" flagship store-- NUO Hotel Beijing has opened grandly, drawing wide attention in China and overseas. As the honorary art advisor of NUO hotel brand, famous artist Mr. Zeng Fanzhi and his team were in charge of the setting and choosing of hotel artworks. And they also invited several artists and architects to specially doing art design for the hotel. Project ROSE is in a relatively dark environment. It is hung above on the public stair platform of basement to mezzanine. The center of ROSE is an attractive flower core lamp which is sometimes bright and sometimes dark, as if it could be touched but also fathomless. The slender rod is made of a new type of environmental protection composite material--steel bamboo, forming a ratio of length and thickness that is beyond understanding, creating an ethereal and charming sense.
This artwork has Archimedes's geometric logic. There is a kind of infinite inward expanding suction force looking from any angle. Maybe this is why the artists name this installation ROSE.
It also amazingly forms the traditional Chinese philosophy "the void", which easily draws people to think of the Eight Diagrams, but with modern and bright features. Moon doors at both sides of the stairs above mezzanine space exactly define the space scope of ROSE. Square and round enhance each other's beauty, which makes us think of "Time Tunnel", as well as the movie Inception.
Therefore, the clearest fantasy may be the best illustration of this artwork.

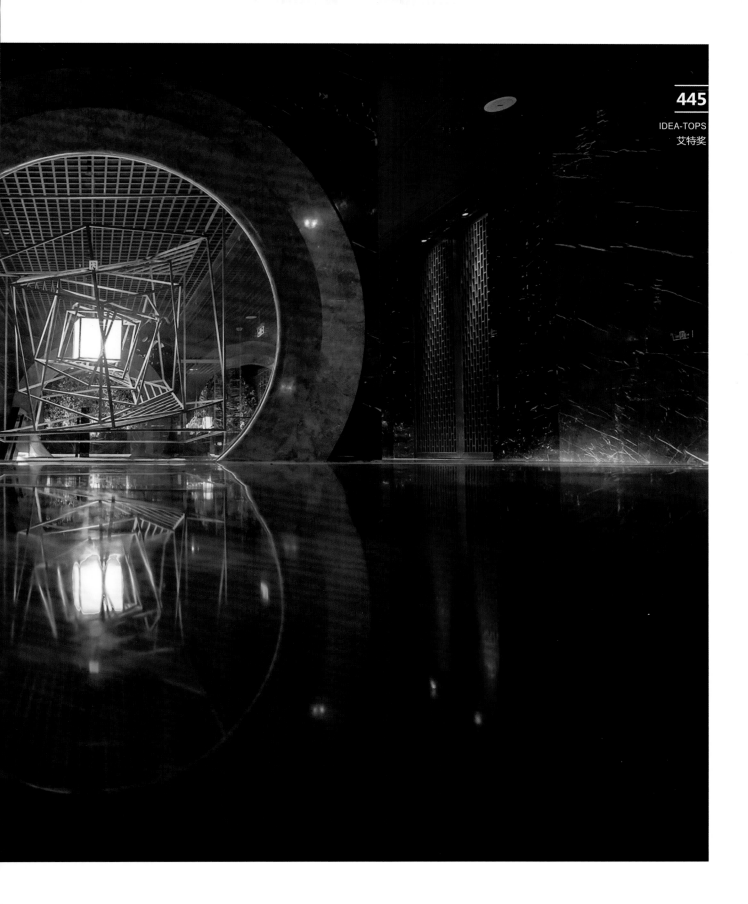

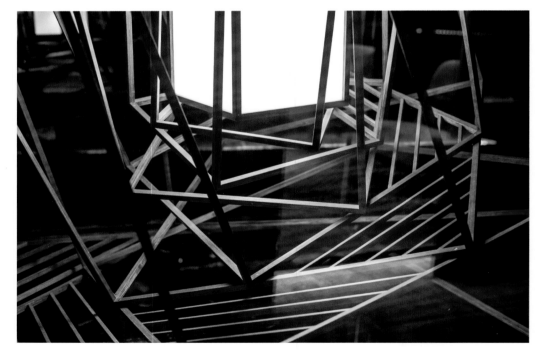

获奖评语

通过不同的线条走势，形成一个完全新的理念。从灯光、造型、方方面面的处理都非常成熟，非常成功。

Through different trends of lines, a completely new idea was formed. The design of lighting, modeling and all other aspects is very mature, very successful.

NOMINATION FOR THE BEST DISPLAY ART DESIGN

最佳陈设艺术设计奖提名奖

Ola-Dele Kuku (比利时/Belgium)

获奖项目/Winning Project

2016威尼斯建筑双年展
尼日利亚馆
Nigerian Pavilion

设计说明/ Design Illustration

2016年第十五届威尼斯建筑双年展

策展人手记：Diminished Capacity（项目名）倾向于用重写历史的志向，去分析一个历史性的交易时刻。由尼日利亚开始，提供一份未公开的说明。在此情况下，重写历史变成必要的进化。对非洲的错误解读将非洲大陆本身转化为一个永远反对不安的国家。那在形式和结构上非洲的本质是什么呢？"非洲不是一个国家！"在此冲突下，首个尼日利亚馆希望探索新方法。冲突是Ola-Dele Kuku周期性的主题之一。这位建筑艺术家将冲突看作当今世界的驱动机制之一，以及在运动中制造改变的工具。"自从创造诞生以来，冲突扮演了一位重要角色，从大爆炸的故事到亚当和夏娃的天堂。"在Kuku先生的实践中，他持续地随着时间重塑表象。在工作中运用画作、装置和物体，他重访了建筑表达方式的基石——平面图、立面图与剖面图——将不安的流动注入严格的形式当中。这种新的作品形式（包括已有的和最近的项目），充分拥抱了一种分析式的社会哲学立场，去面对各种复杂议题，比如资源的消耗、管理和迁移，全球宏观和微观变化，国家能力越发被多种操纵策略的应用所减弱。Camilla Boemio（策展人，评论家）展览人手记：这个策划项目主题为"Diminished capacity"，是当代全球现象"社会-文化冲突"的映射，重点关注"信息/通讯"和"大众传播媒体"的角色。展览是对社交通信和大众传播媒体所产生的摩擦的回应，并直接展示社会和共同价值观一元化趋势的概念。相对于有选择性、有责任感地在特殊性、关注点和监督下进行文章的阐释，大众传播媒体的当代社会学揭示的是全球范围内议题的一致性表达。

15th International Architecture Exhibition Venice Biannale 2016

Curator's note: Diminished Capacity intends to analyze a historical transaction moment with the ambition to rewrite history, starting from Nigeria to provide unpublished interpretations. In this condition, to rewrite history becomes a necessary evolution. The wrong reading of Africa transforms the continent itself into a country poised in perpetual opposition to restlessness; what is its identity in forms and structures? "Africa is not a country!" In that conflict, the first Nigerian Pavilion wants to prospect new methodologies. Conflict is one of the recurrent themes in the work of Ola-Dele Kuku. The architect-artist sees that as one of the driving mechanisms in our world, and as a tool to set change in motion. "Conflict has played a crucial role since the dawn of creation, from the stories of the Big Bang to the paradise of Adam and Eve." Throughout his practice, Ola-Dele Kuku has consistently reshaped representation in a timely challenge. Working with drawing, installation, and objects, he has revisited the mainstays of architectural representational methods – plan, elevation, section – to inject unsettling slippages into their rigorous formalism. This new body of work (consisting of existing and recent projects), fully embraces an analytical socio-philosophical slant that confronts complex issues such as resource depletion and their management, migration, micro and macro global changes, and the diminished capacity of countries amplified by multiple applications of manipulative strategies. Camilla Boemio (curator, critic) Exhibitor's note 'The proposed project theme titled "Diminished capacity", is a reflection of the contemporary global phenomenon of 'Socio-Cultural Conflicts', with specific focus on the role of "Information / Communication' and the 'Mass Media' . The exhibition will be presented as a reaction to the frictions of social communication and the mass media, vis-à-vis the notion of a unitary tendency of society and common values. The contemporary sociology of mass media communication reveals a consistent presentation of agendas rather than reports which are illustrated by selected interest in particularities, focus and oversight' .

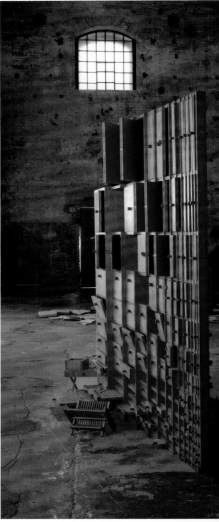

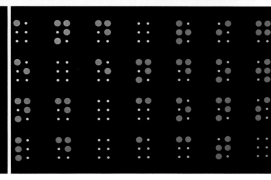

NOMINATION FOR THE BEST DISPLAY ART DESIGN

最佳陈设艺术设计奖提名奖

昊泽
空间

深圳市昊泽空间设计有限公司
（中国深圳）
Shenzhen Horizon Space Design
(Shenzhen, China)

获奖项目/Winning Project

以艺术之名—上海建发璟园
项目A、B户型样板房
软装设计
In the Name of Art

设计说明/ Design Illustration

他们说"要有光"，于是就有了光。
白驹过隙。
一个人的欣赏趣味总会随着年龄而发生变
化，那些人生与作品中的种种不可理喻却感
人至深的疯狂，都是曾在我年少心怀里投下
的巨石，而巨石投在青春的那一瞬所激起的
磅礴巨浪，终于幸或不幸地，在岁月的推移
流转里减弱为似有还无的涟漪。
如果可以把艺术比喻为露在海面之上的冰山
一角的话，那么海面就是创作者本人及其所
生活的那个社会，而冰山隐藏在海面之下的
庞大身躯，亦即那些由毕达哥斯拉、柏拉图、
亚里士多德、圣奥古斯丁等伟大哲人们无声
浸润于社会与艺术之内的诸般奇异哲思。
这，才是令我兴趣盎然之地。

They said, "it needs to have light",
so this place has light.
As time goes, a person's taste
changes with age, those ridiculous
but impressive things I have done
when I was young are like rocks in my
heart. While the big waves set off by

those rocks, fortunately or
unfortunately, become ripples
as time passes.
If we can compare art to a tip
of the iceberg above the sea,
then the surface of the sea is
the society where creators
live in, and those parts which
hide under the sea are the
unusual thoughts of great
philosophers, such as
Pythagoras, Plato, Aristotle
and Saint Augustine, these
thoughts silently influence the
society and art, and this is the
part which makes me really
interested.

NOMINATION FOR THE BEST DISPLAY ART DESIGN

最佳陈设艺术设计奖提名奖

深圳市逸尚东方室内设计有限公司（中国深圳）
Orientalicon Interior Design (Shenzhen, China)

获奖项目/Winning Project

成都中洲锦城湖岸售楼处
Chengdu Zhongzhou The Lake Sales Office

设计说明/ Design Illustration

售楼处位于"天府之国"的成都，整个空间的设计理念是将中式建筑风格和东南亚建筑风格特色相融合，打造一个具国际化水准的城市新境界。在设计之初我们把中国传统文化的精髓与当代时尚潮流加以融合并创新演绎，设计中不单只是纯粹的元素堆砌、复制，而是通过对传统文化的认知，将传承与创新元素相碰撞，符合现代人的审美需求同时来打造富有传统韵味的美好事物。设计中突出空间的层次感，通过"移步借景"表现手法，让空间充分展现中国文化独特的艺术内涵，并将传统特色做到极致，完美诠释空间的意境与惬意。当淳朴而内敛的颜色碰撞尊贵大气跳色，仿佛中国东方元素气息中缓缓流淌着一抹新鲜的血液。将东方意韵引入空间，通过"韵""艺""静""雅"诠释不同空间，营造出"意为境之始，境为意之终"的审美境域。

The real-estate sales center is located in Chengdu, known as "the land of abundance". Our design philosophy is to create a Chinese architecture featuring Southeast Asian construction

elements, in order to offer an internationalized atmosphere to the city. We've combined the essence of traditional Chinese culture with the modern fashion trend not by just copy-and-paste the elements themselves, but appreciate the true value of traditional culture. A brand new aroma will be created to fulfill people's aesthetic demand. To extend the space into multiple-layers scene and present the uniqueness of artistic conception, we adopted the technique of "walk in scene" to bring audience more viewing angles. When an understated color crashes with a delightful flash-color, it feels like a fresh blood is pumped into the veins of Chinese-Eastern elements. With the original allocation of space and arrangement of contrasting colors, the aesthetic of the east is well presented through the four aspects-- "Charm", "Artistry", "Quietness" and "Elegance".

NOMINATION FOR THE BEST DISPLAY ART DESIGN

最佳陈设艺术设计奖提名奖

方印（中国北京）
Fang Yin (Beijing, China)

- -

获奖项目/Winning Project

十八咖啡
Eighteen Coffee

设计说明/ Design Illustration

结合咖啡与餐吧功能融入空间，以复古元素为核心，波普文化元素为道具的国际化品质休闲交流空间；通过空间氛围、色彩的变幻与光线、陈设的运用来体现出空间的主题性，以触及人们放松且印象深刻的状态。

Eighteen Coffee致力于打造年轻、奔放、创新又富有个性的时尚餐饮空间，所以跳脱传统咖啡餐吧的形象束缚，以Pop Art的色彩来赋予空间性格、情绪；椅子、沙发、绿植等每个细节的把控，墙上涂鸦的图案设计等，希望来此消费的客人可以和18空间有所交流共鸣，也就是空间所传达的能量，包间里的艺术画作、细节之处内在的隐微都可以让人们驻足欣喜，让其觉得我们设计的用心，便足矣。

This space combines coffee and dinner together. With retro elements as the key note, it also integrates pop cultural elements, creating an international, leisure communication place. The theme is presented through the use of changeable colors, lighting and decoration, thus people would be impressed by the whole atmosphere and feel relaxed.

Eighteen Coffee aims to create a young, enthusiastic and innovative dining space which is also personalized and fashionable. It gets rid of the traditional impression of a coffee bar, endows characteristic and mood into the space with pop art colors. All the details, including the chairs, sofas, green plants, the graffiti on the walls and the artistic paintings in the room, are meticulously controlled to build resonance between customers and space, to convey the energy of space, to please people and let them know the hard efforts that were made.

466

The
Best
Lighting Environment
Design Award

最佳光环境设计奖

艾特奖

最佳光环境设计奖
THE BEST LIGHTING ENVIRONMENT DESIGN AWARD

INTERNATIONAL
SPACE DESIGN
AWARD

获奖者/Winner
Daan Roosegaarde
(荷兰/Netherlands)

获奖项目/Winning Project
Waterlicht

获奖项目/Winning Project

Waterlicht

设计说明/ Design Illustration

Waterlicht是关于水的力量和诗意的梦想景观。作为一种虚拟洪水，它展示了水在没有人的干预下可以达到多高。提供建造堤坝和创造性思考，创新是荷兰景观DNA中的自然部分，但我们几乎忘记了这个。
Waterlicht是一种值得铭记的充满力量的诗意的体验，通过使用最新LED技术、软件和透镜展示曲线的光。最开始为荷兰地区水务局创造的艺术作品，已经到过阿姆斯特丹、巴黎、斯霍克尔展示。
Waterlicht将继续提升大家对水的关注。它让人们得以体验几乎被遗忘的水的力量和脆弱性。
项目创作人Daan说道："Waterlicht展示了荷兰在虚拟洪水下的样子。在水务系统和我们的历史促进下，我们整个景观都可以看到创新，但我们已经几乎忘记了这些。"这个艺术项目利用最新的LED技术、软件和透镜展示了曲线条的光线。"对于水的关注十分重要，如果没有水务系统，阿姆斯特丹国立博物馆将被水淹没，这就是我们支持这个创新项目的原因。"当地水务局官员Gerhard说道。阿姆斯特丹国立博物馆最近收藏的一幅17世纪的由Jan Asselijn创作的画作——1651年阿姆斯特丹洪水——是临时展览Waterlicht项目的真正原因。这两件作品都反映了荷兰水的历史，以及人、自然、科技之间的互动。这创造了一个极佳的联系。阿姆斯特丹国立博物馆馆长Wim Pijbes说道："画很明白地解释了荷兰的处境，我们一直在海平线以下生活。"荷兰ING公司CEO，Nick Jue说："通过将这个特别的展览向公众开放，我们将古老的画作与现代的"Waterlicht'项目联系起来。"

Waterlicht is the dream landscape about the power and poetry of water. As a virtual flood, it shows how high the water could reach without human intervention. Innovation is a natural part of the DNA of the Dutch landscape through its dikes and creative thinking, yet we almost seem to have forgotten this. Waterlicht is a powerful and poetic experience to remember. Waterlicht consists of wavy lines of light made with the latest LED technology, software and lenses.

Originally created for the Dutch District Water Board Rijn & IJssel, the artwork has now travelled to the Museumplein in Amsterdam, Nuit Blanche in Paris, France and UNESCO Schokland, the Netherlands. Waterlicht will continue its journey to create more water awareness.
Waterlicht lets the visitor experience the almost forgotten power and vulnerability of water.
This inspired Daan Roosegaarde: "Waterlicht shows how the Netherlands look like without waterworks, a virtual flood. Innovation is seen throughout our landscape, pushed by the waterworks and our history, but yet we've almost seem to forgotten this." The artwork consists of wavy lines of light, made with the latest LED technology, software and lenses."Water awareness is crucial, without all our waterworks, the Rijksmuseum for instance could be under water. That is why we support this initiative," says Gerhard van den Top of the local waterboard.
The recent acquisition by the Rijksmuseum of the 17th century painting by Jan Asselijn of the Amsterdam flood in 1651 was the actual reason for the temporary Waterlicht exhibition. Both works reflect on the water history of the Netherlands and the interaction between man, nature and technology. This created a great connection.
Wim Pijbes general director of the Rijksmuseum explains: "The painting explains clearly the dutch situation: we have always lived under the sea-level". Nick Jue, CEO of ING Netherlands: "With this special exhibition open for a wide audience, we combine the old painting with the modern 'Waterlicht' .

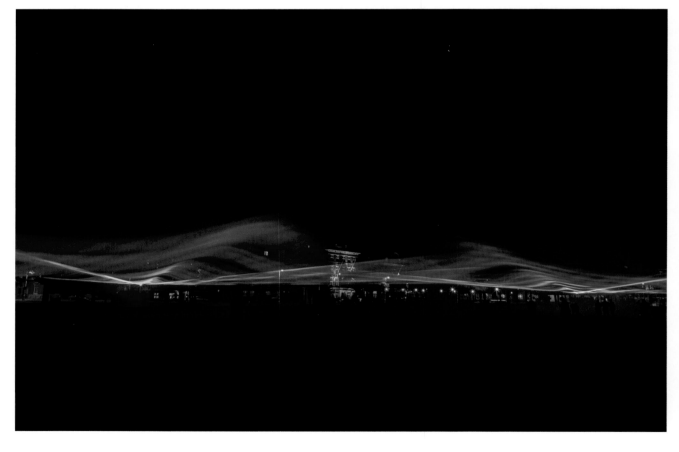

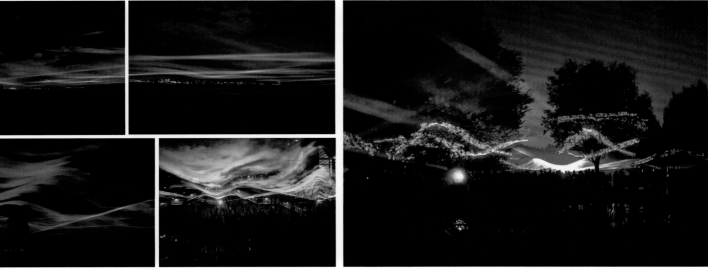

获奖评语

这个方案已经突破了常规的灯光照明的认识，营造了一种梦幻、神秘的氛围，和光环境的营造主题比较契合，给人一种新的体验。

This proposal has broken the conventional understanding of lighting, creating a dreaming and mysterious atmosphere, which comparatively fits the theme of light environment creating, providing a new experience to people.

NOMINATION FOR THE BEST LIGHTING ENVIRONMENT DESIGN

最佳光环境设计奖提名奖

邱春瑞（中国深圳）
Raynon Chiu (Shenzhen, China)

获奖项目/Winning Project

绿景·红树湾一号销售中心
LVGEM - Mangrove No.1 Sales Center

设计说明/ Design Illustration

本案位于深圳市福田区金地一路，地段繁华，环境适宜。现代人的生活情结在空间中得以展示出来，不同的空间带来的是不同的生活品质和生活氛围，人们希望藉由空间这种外在的物质形态给生活带来看不见的、内在的精神。空间的精神如同月亮之于夜空。因有精神，而凸显出空间的美好情怀。

设计师在入口处以室外大面积的水景，试图化解周遭环境的繁杂，而使业主们在被抽离的境遇中感受到禅意和静谧的氛围，这种洗练而大气的造景，无疑是极度出彩的。它以一种成熟而收敛的方式从大环境中抽离出来，使整个空间立于纷繁，而静于内心。

整个大厅区域与洽谈区结合起来，一条长桌贯穿整个空间，以其巨大的仪式感，升华了整个空间的气场，使人们在空旷静谧的空间中，感受到敬意和尊重。同时大厅高层的挑高，浅木色格栅与白色亚克力嵌入其中，在墙面上所形成的巨大体量感，以及垂挂于大厅中心的水晶挂饰与精致的黑色塘池，相互呼应，犹如清泉之水天上来，寓以聚财之意。

在沙盘区域，同样延伸了木色的元素，同时压低了层高，使整个空间的气场更加收敛、聚合，

而使客户在观看沙盘时有一个更加温馨的氛围。

在VIP洽谈区域，于细节上更加的讲究，并配置了挂画，强化了整个空间的文化性，座椅家私的选择也更具人性化和设计感。

在办公区域，这种禅意和静谧的氛围同样得以体现，开放式的设置，兼具娱乐、办公、休闲、洽谈于一体的工作空间，无不体现了设计师前卫的完美主义立场。

整个空间设计，与材质、与光影、与伦理、与空间气场的收放聚散，都颇具大师手笔，不着一处，不留一痕，而将禅意和静谧的内核挥洒得尽致淋漓。

This project is located in Jindi 1st road, Futian District of Shenzhen, a flourishing area which enjoys a favorable environment. The life philosophy of current people is expressed by space, and different space brings us different life quality and living atmosphere, we hope to introduce an invisible and inner spirit to our lives through physical forms. The spirit of a space is like the moon in the sky, it's the spirit that highlights the delightful atmosphere in the space.

Without extravagant decoration, the place is concise and delicate. Moreover, it emits conventional confidence and charm anywhere. This is the most distinguishing feature of the project for this time. The designer uses much waterscape on the entrance of the building, aiming to reduce the noisy feel caused by the surrounding environment. The unique design could help people release their pressure and get the tranquility. There is no doubt that the polished and decent landscaping is extremely brilliant. It's showed through a mature and quiet way.

The hall is linked with the negotiation area, and a long table crosses through the whole space and causes a strong sense of ritual, which highlights the space aura, so that people in the quiet area could feel to be respected. The loft living style with light color wooden gratings and white acrylics has achieved an awesome visual effect. In addition, the hanging crystal decorations and exquisite black pond are perfectly integrated with each other. In traditional Chinese Geomancy Study, it means accumulation of fortune. In sand table area, wooden element also has been adopted. Furthermore, the designer lowered the height of floor so that the whole space is much compact, and customers can enjoy a warm atmosphere.

In the VIP negotiation area, the designer focused more on the details and hanged some pictures on the wall, strengthening the cultural trait. Besides, the seats are well designed in a humanized way.

In working area, the open layout accommodates a comprehensive space for recreation, relaxation, business and negotiation, which shows the designer's perfectionism.

From the design of the whole space, to the flexible usage on material, light and shadow, as well as the space aura, all strongly demonstrate the Zen style and a sense of serenity.

NOMINATION FOR THE BEST LIGHTING ENVIRONMENT DESIGN

最佳光环境设计奖提名奖

柏万军（中国北京）
Bai Wanjun (Beijing, China)

获奖项目/Winning Project

江阴敔山嘉荷精品酒店
Yu-Shan Bay Boutique Hotel

设计说明/ Design Illustration

这个项目是一处旧建筑改造项目，建筑三面玻璃幕墙，一面石材幕墙，整体呈几何化造型。项目定位为精品酒店，给我们塑造建筑外立面形象带来很大的挑战。通过对建筑形体的仔细揣摩，我们找到适合表现建筑的形体，建筑定位为精品酒店的方式，酒店室内设计界定为禅味与现代休闲的感觉。所以我协同室内设计师对光的分布和灯具的承载结构做了一些修订，使得灯光和灯具的分布与室内空间的形式和秩序更加协调，让空间视觉更有秩序感、层次感。

This project is a redesign project of an old building. There are three glass curtain walls and one stone curtain wall, with an overall geometry shape. The project orientation is boutique hotel, which leaves us a big challenge to design the exterior facade of the building. By considering carefully about the shape of the building, we find a proper way to express the building shape, its advantages and the project orientation. The hotel interior design is defined as the feeling of Zen and modern leisure, therefore I coordinate with interior designers and make some adjustments about the light distribution and bearing structure of lamp, so that the distribution of lights and lamps as well as the indoor space form and a sense of order could be more harmonious, which make the space visually more orderly and layered.

NOMINATION FOR THE BEST LIGHTING ENVIRONMENT DESIGN

最佳光环境设计奖提名奖

朱玉晓、 Martin Miller
（中国济南&美国）
Zhu Yuxiao, Martin Miller (Jinan, China&USA)

获奖项目/Winning Project
《茧》——翅的冥想
Cocoon—Meditation of Wings

设计说明/ Design Illustration

瞿颖想要一个独特的空间来放松和与朋友小聚。
我们给出一个放飞的理念：通过自然的生态营造
出无形的翅。
空间概念上，我们试图营造一朵朵轻柔流动的
"云"；内部空间秩序微妙地通过起伏的曲线平
板来创造，这些平板也同时为整个房间创造了多
个独立私密的小空间。透光性片材允许自然光进
入房间，为阳光房增添温暖和亮度。视觉上，当
片材作为分层式的吊架使用时，整个空间会变得
柔软模糊。
与轻柔流动的"云"相融合的是"水滴"般造型
的绿植/鱼缸装置。要创造出放松和冥想的空间
则是水，这是创造这个空间的主要驱动力，它可
以唤起空间内的"柔软性"与"流动性"，创造
出令人舒缓情绪和找到乐趣的流体空间。
硬质的钢丝网被重新塑造，利用其天然的材料特
性来创建出流动的液体"水滴"形态。而"水
滴"像一个符号一般将整个屋子的概念统一。
一个小的射灯安装在绿植装置的底部来引导绿色
藤萝向着钢网的方向攀爬生长；同时也使球形鱼
缸和水创造神奇的光影。
中央的麻绳地毯是精心手工定制的，作为房间内
的流动元素。粗旷的麻绳呈漩涡形绕自身旋转，
如同水波般的圈圈荡漾，地毯则精确排列设计。
所有这些让冥想成为无形的翅，放飞。

Qu Ying wants a special space to relax and meet with her friends.

We offer her a flying concept: through natural ecology to create invisible wings.

In space concept, we try to create gently flowing "clouds". The inner space order is delicately created by fluctuated curve board. These boards also create plenty of independent and secret little space for the room. Transparent sheets allow natural light to enter the room, which adds warmth and brightness. Visually, when sheets are used as layered hanger, the whole space could be softer and vague.There are green plants and fish tank like water drop that combines with softly flowing "clouds". It is water that creates the space of relaxing and meditation, which is the main driving force to create this space. It arouses softness and flowability within the space, creating a flowing space for people to soften mood and find joy.

Hard steel net are being rebuilt, using its natural material feature to create flowing water drop shape.

Meanwhile water drop is like a symbol to unify the concept of the whole house.

A small spotlight is installed at the bottom of the green plant installation to lead green vines to grow toward steel net, at the same time make ball shape fish tank and water create amazing light and shadow.

The central rope carpet is custom-made by hand delicately, as one of flowing elements within the room. Rough rope swirls around its own rotation like a whirlpool, which waves like circles of ripples. The precise arrangement of carpet is completely up to the adjacent rope, vice versa. All of these make meditation become invisible wings, flying.

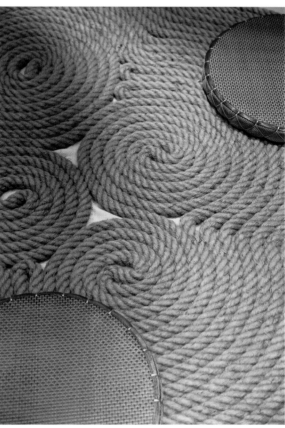

NOMINATION FOR THE BEST LIGHTING ENVIRONMENT DESIGN

最佳光环境设计奖提名奖

范宝太、 曹刚（中国郑州）
Fan Baotai, Cao Gang (Zhengzhou, China)

获奖项目/Winning Project

美祥1969
Meixiang 1969 Wooden Experience Pavilion

设计说明/ Design Illustration

美祥1969木制体验馆，是客户对美祥木制文化的体验地，同时也是静谧的展览空间。室内设计师运用体量不一的木条组团排列，有层次地创造了一个室内的鸟语花香、小树林的静谧优雅空间。

为了呈现出环境的光阴变幻之美，结合室内"小树林"，我们充分考虑在满足功能需求的基础上创造光影变化，神秘的节奏，美轮美奂。为了营造闲暇氛围，采用暖白色，色温在3000K左右的LED射灯放置在林中顶面，同时地面点缀1W小明装射灯。上下交错，明暗对比衬托空间的层次。白，意涵着灵性之白，回归最原始的简洁与纤细，在高纯度的光线下更具清澈感，犹如隔离了室外喧嚣，筑起一隅沉静的空间。

美祥空间划分为几个区域，展示空间除了以上说的灯光外，还设计了符合整体风格的独创瓦片灯，在灯光暗下来时，整个顶面宛如家乡的夜空，美、静、闪烁。整个空间通过智能系统，编辑出欢迎、展览、会客、背景音乐等几种模式，达到配套设备与空间的完美结合。其他区域均用混合的布置灯光的手法，如：洗墙、重点照明等表达出空间的层次。

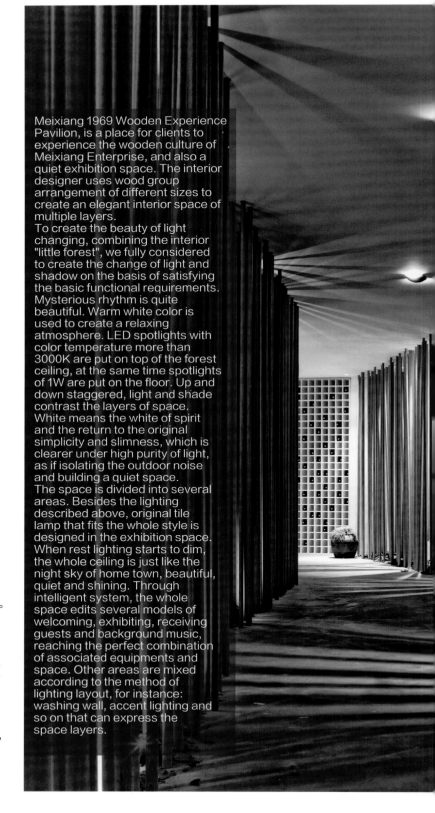

Meixiang 1969 Wooden Experience Pavilion, is a place for clients to experience the wooden culture of Meixiang Enterprise, and also a quiet exhibition space. The interior designer uses wood group arrangement of different sizes to create an elegant interior space of multiple layers.

To create the beauty of light changing, combining the interior "little forest", we fully considered to create the change of light and shadow on the basis of satisfying the basic functional requirements. Mysterious rhythm is quite beautiful. Warm white color is used to create a relaxing atmosphere. LED spotlights with color temperature more than 3000K are put on top of the forest ceiling, at the same time spotlights of 1W are put on the floor. Up and down staggered, light and shade contrast the layers of space. White means the white of spirit and the return to the original simplicity and slimness, which is clearer under high purity of light, as if isolating the outdoor noise and building a quiet space.

The space is divided into several areas. Besides the lighting described above, original tile lamp that fits the whole style is designed in the exhibition space. When rest lighting starts to dim, the whole ceiling is just like the night sky of home town, beautiful, quiet and shining. Through intelligent system, the whole space edits several models of welcoming, exhibiting, receiving guests and background music, reaching the perfect combination of associated equipments and space. Other areas are mixed according to the method of lighting layout, for instance: washing wall, accent lighting and so on that can express the space layers.

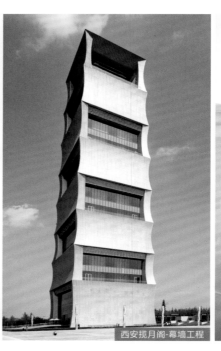

华宏诚
金属屏风 | 隔断 | 花格

金属隔断 设计定制
中式欧式 复古现代

华宏诚金属制品有限公司
电话：0757-88748788
手机：136 9515 1145
网址：www.hhcbxg.com.cn
广东省佛山市禅城区澜石

扫一扫上面的二维码图案，关注更多信息

隐庐

T : 0755-26823397
E : hsdesigner@126.com
地址：深圳市南山区南海意库2栋413房

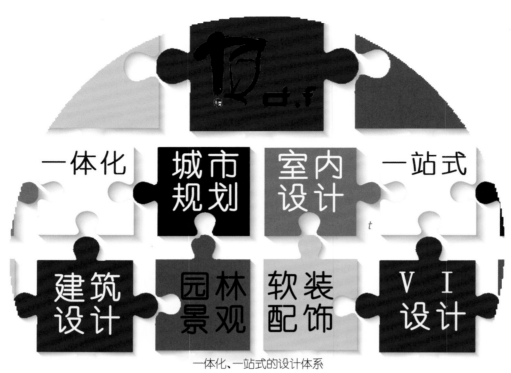

一体化　城市规划　室内设计　一站式

建筑设计　园林景观　软装配饰　VI设计

一体化、一站式的设计体系
主要业务囊括了城市规划、建筑设计、园林景观、室内设计以及软装配饰和VI设计

林伟权　主笔设计师　张津铭　主笔设计师　尼奇·波帕尔　主笔设计师　主创规划师　邓学杰

大梵设计创始人　宋亚东　吴海峰　大梵设计创始人

选择深圳市大梵设计集团可以让您更快更好地
体验到优质设计带来的美好生活。

深圳市大梵设计集团联系电话：0755-36607005

dafan 传真：0755-86099086

地　址：深圳市宝安区宝源路F518
时尚创意园 F1栋103-104，201-202

盛堂 SAINTANG

盛堂设计顾问有限公司由盛思民先生创办，专业为全国范围内的各类商业空间、文化空间、住宅空间、整体软装提供完整的设计及顾问服务。

我们建立了设计、工程技术、配饰服务以及高端设计产品供货的链条服务系统，能够满足客户对全面设计服务的需求，在设计服务方面为客户提供便利，并有效控制最终设计项目的实施品质。

团队主创人员皆有多年服务国内高端项目经验，与客户建立了良好合作关系，愿意长期与信任我们的新老客户共同成长和发展。

盛思民

深圳市盛堂设计顾问有限公司 创办人 设计总监
江西设计·深圳　执行会长
景德镇学院客座讲师 毕业生实践导师
深圳市室内设计师协会理事会 副秘书长
深圳市陈设艺术协会 常务理事
龙承奖 评委副主席

2016年　被授予"大中华区酒店会所十佳设计师"称号
2014年　被授予"大中华区别墅十佳设计师"称号
2014年　被授予"深港杰出设计师"称号
2010年　第五届海峡两岸四地室内设计优秀设计师
2015年　受邀参加CCTV2交换空间栏目
2016年　作品以"空间榜样"项目在CCTV2播放
作品曾在CCTV华人频道《中国好设计栏目》、新浪、腾讯、深圳市装饰网、设计本、A963、现代装饰、《南方都市报》、饰家爱家客等多家媒体刊登。

深圳市盛堂設計顧问有限公司
Shenzhen Saintang Design & consultant Ltd.

深圳市 南山区 侨香路 香年广场A座 2001室
Room 2001,Building A,Xiangnian Plaza
Qiaoxiang Road,Nanshan District
Shenzhen City,Chiana
服务热线：4006-050-330
电话Tel：0755-28169915

GABRIEL
圣嘉佰利

高端欧美整体家居供应商

高端LED·灯饰·家具·饰品·风扇一站式品牌服务商

圣嘉佰利品牌简介：

    圣嘉佰利创立于2005年，是一家集灯饰·LED·灯扇·家具·饰品于一体的高端欧美整体家居创意品牌，有十多年的专业产品设计、生产工艺制作及品牌营销管理经验。圣嘉佰利始终关注人们对于高品质生活的追求，以让人们享受欧美高品质生活为愿景，融欧式的典雅、英伦的贵气、美式的古朴、中式的文化于一体，用现代工艺传承古典艺术，中西合璧，透古通今。

    古罗马的雕像、意大利的美术手工、欧式古堡建筑、美式乡村艺术、国内宫廷园林……圣嘉佰利的每一款产品都深浸艺术的魅力，彰显着文化的内涵。让经典得以传承，将艺术基因植入居室生活，把怀旧、爱、温馨、浪漫、品味、舒适等情感要素融入品牌文化内核，从而唤起人们内心深处的渴望与共鸣。

    经过10多年的发展，圣嘉佰利在国内已建立起300多家品牌连锁专卖店渠道销售网络。集欧美风格灯饰、LED智能照明、灯扇、家具、饰品整体家居配套产品数千款，同时还可根据客户需求提供非标定制服务，能极大程度地满足国内不同消费群体的多种需求。

    圣嘉佰利秉承"经营严谨、设计考究、客户至上、互利共赢"的经营理念，以"敬业感恩、求实创新、高效共赢"的企业核心价值观，想行业所想，急客户所急。以恒稳的品质，大众化的价格，集装饰性照明、功能性照明和系列软装配套产品构建"大而全、全而精"的品牌发展格局。

    装饰生活，照亮梦想。圣嘉佰利以爱和光明庇佑每个家庭，为每个家庭缔造温馨舒适的居室生活空间。

圣嘉佰利官网

微信关注"圣嘉佰利"获取更多内容

展厅地址：广东省东莞市大朗镇新马莲工业区圣嘉佰利运营中心

服务热线：400-8298-169　　网址：www.chinagabriel.com

效果图

罗湖北 云创城 生活家

传承康利匠心品质，筑就企业形象封面
24万平方米地标级产城综合体，刷新科技办公环境标准

总部商务，全球热租

※ 双地铁物业	片区罕有产权到户	6米层高阔绰空间	总部级大平面办公	52部OTIS电梯
罗湖北东进规划双核	资产价值恒久	生活无限创意	片区首发	4层阔绰智能停车场

KANGLI CITY
康利城

0755´
8996 1999
深圳' 平吉大道66号　开发商:深圳康利置地有限公司

整合推广: EXC 例外传播　│建筑设计:柏涛建筑　│销售代理:世联行

扫一扫 更多惊喜!

HUGE ROCK DESIGN
盘石创意·定制级设计

HUGE ROCK DESIGN
HTTP://WWW.HRDSZ.COM

文粒设计事务所
WU WENLI DESIGN FIRM

T E L : 5 - 2 2 2 0 0 0 0 9
E : 7 7 @ 1 6 3 . C O M

址 ： 深 圳 南 徐 香 路 与 深 云 路 交 汇 香 年 广 场 B座 2 0 1

图书在版编目（ＣＩＰ）数据

第七届艾特奖获奖作品年鉴/国际空间设计大奖艾特奖
组委会编著.--长沙:湖南科学技术出版社,2017.9
ISBN 978-7-5357-9432-1

Ⅰ.①第… Ⅱ.①国… Ⅲ.①室内装饰设计—世界—图集
Ⅳ.①TU238-64

中国版本图书馆CIP数据核字(2017)第202436号

第七届艾特奖获奖作品年鉴

编　　著：国际空间设计大奖艾特奖组委会
责任编辑：缪峥嵘
总　策　划：深圳市东方辉煌文化传播有限公司
统筹策划：赵庆祥
出版发行：湖南科学技术出版社
社　　址：长沙市湘雅路276号
　　　　　http://www.hnstp.com
邮购联系：本社直销科　0731-84375808
印　　刷：深圳市祥龙印刷有限公司
　　　　　（印装质量问题请直接与本厂联系）
厂　　址：深圳市龙岗区平湖辅城坳岐新一路48号
邮　　编：518100
出版日期：2017年9月第1版第1次
开　　本：889mm×1194mm 1/16
印　　张：31.25
书　　号：ISBN 978-7-5357-9432-1
定　　价：598.00元